BLACKBERRY HACKS™

Other resources from O'Reilly

Related titles

Hardware Hacking Projects for Geeks

PC Hardware in a Nutshell

PC Hardware Annoyances

Treo Fan Book

PC Hardware Buyer's Guide

Palm and Treo Hacks

Building the Perfect PC

Home Hacking Projects for Geeks

Podcasting Hacks

Hacks Series Home

hacks.oreilly.com is a community site for developers and power users of all stripes. Readers learn from each other as they share their favorite tips and tools for Mac OS X, Linux, Google, Windows XP, and more.

oreilly.com

oreilly.com is more than a complete catalog of O'Reilly books. You'll also find links to news, events, articles, weblogs, sample chapters, and code examples.

oreillynet.com is the essential portal for developers interested in open and emerging technologies, including new platforms, programming languages, and operating systems.

Conferences

O'Reilly brings diverse innovators together to nurture the ideas that spark revolutionary industries. We specialize in documenting the latest tools and systems, translating the innovator's knowledge into useful skills for those in the trenches. Visit *conferences.oreilly.com* for our upcoming events.

Safari Bookshelf (*safari.oreilly.com*) is the premier online reference library for programmers and IT professionals. Conduct searches across more than 1,000 books. Subscribers can zero in on answers to time-critical questions in a matter of seconds. Read the books on your Bookshelf from cover to cover or simply flip to the page you need. Try it today for free.

BLACKBERRY
HACKS™

Dave Mabe

O'REILLY®

Beijing · Cambridge · Farnham · Köln · Paris · Sebastopol · Taipei · Tokyo

BlackBerry Hacks™

by Dave Mabe

Published by O'Reilly Media, Inc., 1005 Gravenstein Highway North, Sebastopol, CA 95472.

O'Reilly books may be purchased for educational, business, or sales promotional use. Online editions are also available for most titles (*safari.oreilly.com*). For more information, contact our corporate/institutional sales department: (800) 998-9938 or *corporate@oreilly.com*.

Editor:	Brian Jepson	**Production Editor:**	Jamie Peppard
Series Editor:	Rael Dornfest	**Cover Designer:**	Marcia Friedman
Executive Editor:	Dale Dougherty	**Interior Designer:**	David Futato

Printing History:

October 2005:	First Edition.

 This book uses RepKover™, a durable and flexible lay-flat binding.

ISBN: 0-596-10115-5
[M]

Contents

Credits

About the Author

Dave Mabe (*http://dave.runningland.com*) is an accomplished and largely self-taught engineer and now a writer who strives to create a simple, elegant solution to a complex problem. Dave has worked at AT&T in the communications industry for eight years and has worked with BlackBerry devices for almost five. Always looking to save a few keystrokes and mouse clicks, Dave has automated countless business processes for Fortune 500 companies. He is the kind of person who would rather spend several hours inventing an automated solution than spend a few monotonous moments each day performing a menial task. He fancies himself a proficient Perl programmer and would love to see Perl running on a BlackBerry one day. Dave enjoys tinkering with the latest gadgets and software, both closed and open source. He has authored and contributed to several technology-related publications.

When not working on computer-oriented stuff, you're likely to see Dave running on the roads and trails (mostly trails!) of Chapel Hill, where he graduated with a BS in Mathematical Sciences, specializing in Operations Research. Although quite a bit slower than the glory days of college running at UNC, he can still leg out a decent 10K. Dave lives in Chapel Hill, North Carolina with his three energetic daughters, Sarah Jane, Rosemary, and Lizzie. Dave's running is easily overshadowed by that of his wife, Joan Nesbit Mabe (*http://www.runningland.com*), who is a talented runner, writer, and public speaker, who represented the United States in the 1996 Olympic Games in Atlanta. Not being the techie type, she is still not exactly sure what Dave does for a living, but enjoys her Perl-automated cup of coffee each morning nonetheless.

Contributors

The following people contributed their time, writing, hacks, and knowledge toward the writing of this book:

- By day, Phil Bogle is cofounder and chief technology officer of Jobster (*http://www.jobster.com*), a web service that helps companies target and connect with passive jobseekers, while enabling the best candidates to stand out from the pack and get noticed. By night, Phil is a gentleman coder and author of several free BlackBerry applications, including Berry 411 local search. He has a weblog at *http://www.thebogles.com*.

- Paul Dumais is currently vice president of product development at Idokorro Mobile, the leader in mobile network administration software. Paul is the creative and technical force behind Idokorro's products, Mobile Admin and Mobile SSH. Prior to Idokorro, Paul created a world-class bug tracking software that was acquired by MKS Inc and branded Change Integrity. Paul has also held positions as a Java developer at Sun and Corel.

- Jeff Greenhut is founder and president of somedevelopers inc. (*http://www.somedevelopers.com*), a provider of mobile applications for the BlackBerry. After graduating from the State University of New York at Binghamton with a National Merit Scholarship, Jeff has built flight simulators, storage solutions, and accounting software. When not working with computers, Jeff can typically be found working with computers, playing ice hockey, or floating on the lake with BlackBerry in hand. Jeff is married, has two young girls, and lives in the Atlanta, Georgia area.

- Brian Jepson is an O'Reilly editor, programmer, and coauthor of *Mac OS X Tiger for Unix Geeks* and *Linux Unwired*. He's also a volunteer systems administrator and all-around geek for AS220 (*http://www.as220.org*), a nonprofit arts center in Providence, Rhode Island. AS220 gives Rhode Island artists uncensored and unjuried forums for their work. These forums include galleries, performance space, and publications. Brian sees to it that technology, especially free software, supports that mission.

- Shari Kornberg is a software engineer at AT&T, responsible for engineering and support of the Exchange email (50,000 users) and BlackBerry platforms. She designed, architected, and implemented AT&T's BlackBerry infrastructure. Shari was also the technical lead for the recent migration effort to Exchange 2003. Shari began her IT career at AT&T over 25 years ago and has a BA in computer science from Queens College, City University of New York. In her spare time, she enjoys music, theater, travel, and spending time with her three sons, Jeffrey, Alex, and David. You can reach her at *sharimk@gmail.com*.

- Jason Lam is a wireless and open source developer enthusiast who enjoys creating synergy and sharing knowledge in the software development world. As senior wireless developer at QuoteMedia Inc (*http://www.quotemedia.com*) he leads the development of Quotestream Wireless, a financial wireless streaming application. In his spare time, he contributes to open source projects and online tutorials/articles. One of his significant contributions is his open source book *J2ME & Gaming*, (*http://j2megamingbook.sourceforge.net*). To learn more about him, visit his personal site at *http://www.jasonlam604.com*.

- R. Emory Lundberg lives in Providence, Rhode Island, with his wife, Elizabeth, and a chubby calico cat named Echo(1). By day, he applies ninja tactics to errant packets for VeriSign's excellent Managed Security Services team, and by night, he writes about mobile technology and tinkers with gadgets. He will one day be survived by his motor scooters and an exquisite collection of mobile phones.

- Mark Rejhon works at Idokorro Inc. (*http://www.idokorro.com*) and is a moderater of BlackBerry Forums at *http://www.BlackBerryForums.com*. He has a personal web site at *http://www.marky.com*.

Acknowledgments

This book couldn't have been written without the generosity of the contributors and the helpful guidance of my wonderful editor, Brian Jepson. Brian's gentle supervision was reassuring through the entire process and he always provided a quick answer when I had a question. Thanks, Brian!

I have to thank my wife and best friend, Joan, and my daughters who tolerated (and sometimes celebrated?) my frequent absence as this book was formed. They are as much a part of this book as I am, and I will forever be grateful for their sacrifice.

I'd also like to thank my extended family and friends, especially my mom and dad for raising me and remaining two of my closest friends—have fun with those new BlackBerrys!

Many of the hacks couldn't have been conceived or written without the generosity of Loren Beckerman of Cingular Wireless who stepped up to the plate to provide a BlackBerry and best-in-class wireless service throughout the writing of this book.

Thanks to Victoria Berry, Rachael Babcock, and Christopher Herstine (all from Research In Motion) who contributed resources, guidance, and contacts toward this project.

I'd be remiss if I didn't acknowledge the significance of Robert Kern of TIPS Technical Publishing, Inc. (*http://www.technicalpublishing.com/*). His tireless effort helped shape this project from the beginning. Bob, I hope to see you on the trails soon!

Preface

The BlackBerry has come from relative obscurity to omnipresence in a very short amount of time. How was Research In Motion, a relatively small Canadian company, able to steal the thunder from the dominant players in the handheld arena and revive what was becoming a lackluster market? The Internet's original killer app: email. For years no handheld could match the BlackBerry's knack for email. From the small but usable QWERTY keyboard to the push technology that delivers email to a device as it's received, BlackBerry has reached a level to which even the largest players in the industry must scramble to catch up.

Meanwhile, RIM is intent on conquering other aspects of the mobile user experience such as web browsing and secure corporate data access. RIM continues to pour improvements not only into the device itself, but the entire BlackBerry platform. Everyone from the end user to the BlackBerry administrator to the developer will find conveniences in the BlackBerry that don't exist on other platforms. In this book, you'll find clever uses for some of these enhancements and new tricks for using some features that have been there from the beginning.

Why BlackBerry Hacks?

The term *hacking* has a bad reputation in the press. They use it to refer to those who break into systems or wreak havoc with computers as their weapon. Among people who write code, though, the term *hack* refers to a "quick-and-dirty" solution to a problem, or a clever way to get something done. And the term *hacker* is taken very much as a compliment, referring to someone as being *creative*, having the technical chops to get things done. The Hacks series is an attempt to reclaim the word, document the good ways people are hacking, and pass the hacker ethic of creative participation on to the uninitiated. Seeing how others approach systems and problems is often the quickest way to learn about a new technology.

The BlackBerry is ripe with opportunities for the hacker in all of us. Just the size of the keyboard and screen required RIM to build hacks into the device to make it usable for the average person. There are a slew of hacks for the beginner, but even the most advanced power user will find juicy nuggets to incorporate into daily use.

How to Use This Book

You can read this book from cover to cover if you like, but each hack stands on its own, so feel free to browse and jump to the different sections that interest you most. If there's a prerequisite you need to know about, a cross-reference will guide you to the right hack.

How This Book Is Organized

The hacks in this book vary by type and sometimes by audience. As such, it is divided into nine chapters:

Chapter 1, *Using Your BlackBerry*
> This chapter focuses on getting around in your BlackBerry. There are hacks to change programs, do things quickly, hide rarely used icons, and other useful tricks. Use this chapter to make your BlackBerry your own.

Chapter 2, *Email*
> Use this chapter to become a black belt in the BlackBerry's bread and butter. The hacks in this chapter will help you use the device to interact with the application that even your grandmother now uses.

Chapter 3, *Games*
> Where would a computing platform be without its games? Kill some time in the airport waiting for your next flight with the hacks in this chapter.

Chapter 4, *The Internet and Other Networks*
> Interact with the web services that make the Internet what it is today. The Web has transformed handhelds from simple personal digital assistants to truly mobile computers.

Chapter 5, *Free Programs*
> Where there's an operating system, there are sure to be programs available for it. The applications highlighted in this chapter are free for the taking.

Chapter 6, *Shareware Apps*
> If you're willing to drop a little coin, you can use your BlackBerry to perform a variety of functions from multimedia hacks to spellchecking. This chapter features some of the coolest third-party applications out there.

Chapter 7, *BES Administration*
> BlackBerry has become a mission-critical service in the eyes of the corporate executives that use it. BlackBerry administrators will find helpful techniques to help them keep their sanity.

Chapter 8, *The Web and MDS*
> BlackBerry's Mobile Data Service gives users secure access to data from the Internet as well as the intranet. Use this chapter to turn your tired intranet site into a mobile, push-enabled application optimized for the BlackBerry.

Chapter 9, *Application Development*
> Use this chapter as a guide for jumping into BlackBerry development with two feet. These aren't just simple "Hello World" hacks—they are high octane techniques that show even beginning developers how to do powerful things with the BlackBerry.

Where to Learn More

This field is changing at a quick pace. There are some excellent sites to visit to keep abreast of the latest happenings in the world of BlackBerry:

- The BlackBerry Forums (*http://www.blackberryforums.com/*) is a third-party forum site where users exchange questions and answers about BlackBerry. If you run into a problem, there's a good chance someone else has, too.

- The RIM section of Howard Forums (*http://www.howardforums.com/*) is a frequently used forum for exchanging tips and tricks for BlackBerry usage.

- BlackBerry Cool (*http://www.blackberrycool.com*) is an excellent blog that regularly posts software reviews, general news, and tips for BlackBerry users.

- BBHub (*http://www.bbhub.com*) is another heavily posted weblog that highlights new software and news on the BlackBerry.

- BlackBerry.com (*http://www.blackberry.com*) is the official site of BlackBerry. You'll find the latest handhelds, press releases, webcasts, and insight right from the source.

- The BlackBerry Developer Journal (*http://www.blackberry.com/developers/journal/index.shtml*) is a free monthly email newsletter for developers that provides the latest techniques, APIs, and news for BlackBerry developers.

- The BlackBerry developer forums (*http://www.blackberry.com/developers/forum/*) allow developers to interact and ask questions on anything related to the BlackBerry. There is currently a Java forum and a Browser forum.

- The BlackBerry Blog (*http://blackberryblog.com/*) is another blog to monitor the latest BlackBerry happenings.

- The BlackBerry tag on del.icio.us (*http://del.icio.us/tag/blackberry*) is a nice collection of the latest URLs that alpha geek BlackBerry users are bookmarking.

Conventions

The following is a list of the typographical conventions used in this book:

Italics
> Used to indicate URLs, filenames, filename extensions, and directory/folder names. For example, a path in the filesystem will appear as *C:\Temp*.

`Constant width`
> Used to show code examples, the contents of files, and console output, as well as the names of variables, commands, and other code excerpts.

`Constant width bold`
> Used to highlight portions of code, typically new additions to old code. It also indicates text that you should type in literally.

`Constant width italic`
> Used in code examples and tables to show sample text to be replaced with your own values.

Gray type
> Used to indicate a cross-reference within the text.

Alt-X
> Press and hold Alt and tap X, in that order.

Alt-X,Y,Z
> Press and hold Alt, and then type X, Y, and Z, in that order. Release Alt when you are done typing.

You should pay special attention to notes set apart from the text with the following icons:

This is a tip, suggestion, or general note. It contains useful supplementary information about the topic at hand.

This is a warning or note of caution, often indicating that your money or your privacy might be at risk.

The thermometer icons, found next to each hack, indicate the relative complexity of the hack:

beginner   moderate   expert

Using Code Examples

This book is here to help you get your job done. In general, you may use the code in this book in your programs and documentation. You do not need to contact us for permission unless you're reproducing a significant portion of the code. For example, writing a program that uses several chunks of code from this book does not require permission. Selling or distributing a CD-ROM of examples from O'Reilly books *does* require permission. Answering a question by citing this book and quoting example code does not require permission. Incorporating a significant amount of example code from this book into your product's documentation *does* require permission.

We appreciate, but do not require, attribution. An attribution usually includes the title, author, publisher, and ISBN. For example: "*BlackBerry Hacks* by Dave Mabe. Copyright 2006 O'Reilly Media, Inc., 0-596-10115-5."

If you feel your use of code examples falls outside fair use or the permission given above, feel free to contact us at *permissions@oreilly.com*.

How to Contact Us

We have tested and verified the information in this book to the best of our ability, but you may find that features have changed (or even that we have made mistakes!). As a reader of this book, you can help us to improve future editions by sending us your feedback. Please let us know about any errors, inaccuracies, bugs, misleading or confusing statements, and typos that you find anywhere in this book.

Please also let us know what we can do to make this book more useful to you. We take your comments seriously and will try to incorporate reasonable suggestions into future editions. You can write to us at:

O'Reilly Media, Inc.
1005 Gravenstein Highway North
Sebastopol, CA 95472
(800) 998-9938 (in the United States or Canada)
(707) 829-0515 (international/local)
(707) 829-0104 (fax)

To ask technical questions or to comment on the book, send email to:

bookquestions@oreilly.com

The web site for *BlackBerry Hacks* lists examples, errata, and plans for future editions. You can find this page at:

http://www.oreilly.com/catalog/blackberryhks

For more information about this book and others, see the O'Reilly web site:

http://www.oreilly.com

Safari Enabled

 When you see a Safari® Enabled icon on the cover of your favorite technology book, that means the book is available online through the O'Reilly Network Safari Bookshelf.

Safari offers a solution that's better than e-books. It's a virtual library that lets you easily search thousands of top tech books, cut and paste code samples, download chapters, and find quick answers when you need the most accurate, current information. Try it for free at *http://safari.oreilly.com*.

Got a Hack?

To explore Hacks books online or to contribute a hack for future titles, visit:

http://hacks.oreilly.com

CHAPTER ONE

Using Your BlackBerry
Hacks 1–21

If you just got your first BlackBerry, you've probably already figured out the basics. The nice thing about the BlackBerry is that is has something for every type of user—from "wet behind the ears" newbies to the most advanced alpha-geeks. This chapter uncovers some of the tricks you may not have known your device was capable of. New users will be happy to know what's just below the surface: a clipboard [Hack #2], multitasking [Hack #6], and wireless calendaring [Hack #4]. The hackers in the crowd might like to display the signal strength in decibels instead of bars [Hack #17], use your computer as a wireless headset [Hack #16], or get mobile Internet access on your computer [Hack #9]. Dive right in—there's something for everyone.

HACK #1 Choose a Data Plan

You may know how to choose a voice plan, but what type of data plan do you need?

Mobile phones have been around long enough for people to know the ins and outs of a voice plan. By now, most of us realize what type of mobile phone user we are, how many minutes we're likely to use in a month, and we can pick an appropriate voice plan and realize the implications.

But what about the data plan? How much data are you likely to use with your BlackBerry? What are the factors that should contribute to this decision?

BlackBerry's Wireless Usage

One of the reasons BlackBerry has raced on the scene and become so popular is its "push" technology. Access to email from a mobile device has been around for quite a while, but it's always been a "pull" technology. That is, when you wanted to check your email on your mobile phone, you access a mail application and then poll your server to see if any new mail has arrived. This process is usually slow due to network latency and hardware of small devices.

Research In Motion (RIM), the company that makes the BlackBerry, stepped in and developed a push technology that delivers new emails to your mobile device. This allows you to set up alerts and even be proactively notified when you receive certain email messages [Hack #30].

While this approach to mobile email has been wildly successful, your device's data usage using this model will be far greater since you are not in control of when to use the data connection. This is analogous to accessing the Internet with broadband versus dial-up. Your BlackBerry is like broadband—it is on all the time.

Other Factors to Consider

If you know you'll be using your device with your company's BlackBerry Enterprise Server (BES), your data usage could be significantly lower. Introduced with BES 4.0 is the BlackBerry Router, a piece of software that is designed to know at any given time the best (and cheapest) route to communicate to your device. When you're in the office and cradled, the BlackBerry Router can detect this condition and use the local IP network to send data to your device through the USB cable on your PC instead of "over the air."

This factor will become even more important when RIM produces a much anticipated GSM/WiFi device. This device will theoretically be able to determine the optimal connection depending on your proximity to an 802.11 WiFi hotspot. When in WiFi range, it will be able to "roam to WiFi" and communicate exclusively over its 802.11 adapter. Only when it's out of WiFi range will it use your carrier's GPRS data network. Your company's BlackBerry Router will be able to detect your connectivity and send data through the appropriate route.

The Case for the Unlimited Data Plan

Unless you have good reasons not to, you should go with an unlimited data plan. First of all, the price is typically not much more than the plans that are billed per megabyte (for example, at the time of this writing, Cingular's 4-MB plan was $34.99 a month versus $44.99 for their unlimited data plan). Second, because data is pushed to your device, you have little control over the amount of data that gets sent.

You get plenty of email as it is. Add to that the increasing amount of spam and "joe-jobs" that can occur and it won't take long to go over your monthly allotment. As an administrator who sets up jobs to alert via email, I've certainly been "spammed" by a job that went haywire.

As you discover the third-party software to download and install over the air, the multitude of sites that can be accessed via the BlackBerry Browser, and as the sheer volume of email continues to escalate, you'll be more comfortable with an unlimited data plan.

Do You Have to Have a Data Plan at All?

Actually, with most carriers, the data plan and the voice plan are optional. There are some users who use the device without a calling plan and use only the data features of the device. There are others (fewer, for sure) who use the BlackBerry as a phone without a data plan. The BlackBerry loses a significant amount of its usefulness when you don't have a voice plan and is certainly a lot less fun without a data plan. I won't even address the situation in which one has a BlackBerry device with no voice or data plan—that's just silly.

> Although the voice plan is optional for most carriers, for technical reasons, they still have to actually assign you a telephone number even if you don't use it. Be careful, though—when you choose to have no voice plan, you can still use the phone to make calls. Those calls you make cost an arm and a leg, too. So, when they say you have no voice plan, they really mean "The Worst Voice Plan Ever."

HACK
#2

Cut and Paste Text

Did you know the BlackBerry has a clipboard? Use it to save tons of keystrokes.

As convenient as RIM has made typing on your handheld, it's still—let's face it—typing on a handheld. Any time you can get away without typing text on your device or using shortcuts **[Hack #7]**, you should. This will allow you to be more efficient and crank out more messages from your device with less effort.

One of the most useful ways to save keystrokes is by using BlackBerry's built-in clipboard to copy and paste text from one place to another on your device. Although this is a trivial task on an actual computer, it is a little more difficult on your BlackBerry. Committing these steps to memory now will definitely go a long way in your becoming the ultimate BlackBerry road warrior.

Make Your Selection and Copy It

Once you have identified the text you would like to copy, you need to select it, just like on your computer. Position the cursor to the first character in the text you'd like to select. Remember that holding down the Alt key while scrolling with the trackwheel moves your cursor horizontally. Once you've positioned the cursor on the first character of your selection, hold down the Alt key, click the trackwheel once, then release the Alt key.

This puts the BlackBerry in *selection mode*. Notice the cursor covers only the bottom half of a character when in this mode as shown in Figure 1-1.

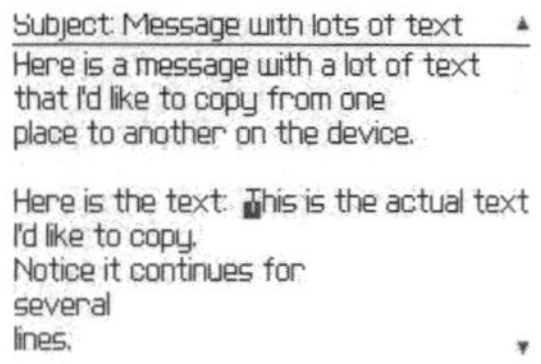

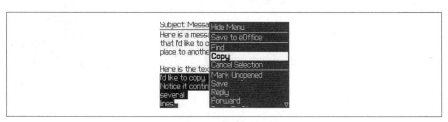

Figure 1-1. Selection mode

Use the trackwheel to select the text you'd like to copy to the clipboard by scrolling up and down. Notice that entire lines are selected when you scroll. Hold down the Alt key while you scroll with the trackwheel to move horizontally on the current line one character at a time.

Once you've selected all the text you'd like to copy, click the trackwheel once to bring up the menu. Notice that when you are in "selection mode" the menu will have a couple extra items with the default selection being Copy. Choose Copy to send the selected text to the clipboard (see Figure 1-2). Notice after you copy text you are no longer in selection mode.

Figure 1-2. The Copy item on the menu in selection mode

Paste the Copied Text

Once you've copied the text to the clipboard, go to the area where you'd like your selected text to appear—most likely the body of a message you are composing. When you have placed the cursor in the proper position, click the trackwheel once to bring up the menu. Notice in Figure 1-3 that when you have text on the clipboard and your cursor is in an editable field, the menu contains a paste function and it's the default selection.

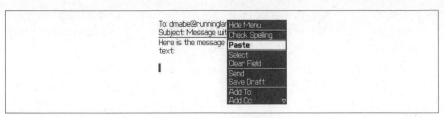

Figure 1-3. The Paste option on the menu

Select the Paste option on the menu to copy the text to the desired position, and voilà! The text appears as shown in Figure 1-4. Once you memorize this technique, you'll probably want to use the keyboard shortcut for pasting (hold down the Caps Lock key and click the trackwheel).

To: dmabe@runningland.com
Subject: Message with Pasted Text
Here is the message with the pasted text:

This is the actual text I'd like to copy. Notice it continues for several lines.

Figure 1-4. The pasted text

Just like in Windows or Mac OS, the selected text remains on the clipboard after it's been pasted until it's replaced with another selection. Also, the clipboard lives in the operating system so it is available across applications. This lets you copy text from a web site via the BlackBerry Browser into a new message in the Mail application.

HACK #3 Take the Easy Way Out with Shortcuts

Cut the course with shortcuts—perhaps the most powerful and underused feature of the device.

The BlackBerry has a very rich system of using shortcuts and hotkeys to quickly navigate inside of applications, and to launch applications from the Home screen. This hack is based largely on the compilation of hotkeys and shortcuts that Mark Rejhon has been maintaining on the very useful Black-Berry Forums web site (*http://www.blackberryforums.com/*). Mark maintains several FAQs that are of interest to BlackBerry newbies and veterans.

To use application launch shortcuts, you may need to disable the ability to dial from the Home screen [Hack #18] in the Phone application's options. This will mean you can't just start dialing numbers when you pick up an idle

BlackBerry, but you can quickly tap the top button on your BlackBerry to put you into the Phone application, and thus be ready for number input there as well.

The Alt key is on the left side of your keyboard. It is just below the A and to the left of the Z. The Shift key is the same as the zero on your keyboard, or it may also be on the right of the keyboard, left of the Power Button and right of the spacebar.

When I say Alt-M, I mean press and hold Alt and tap M, in that order.

When I say Alt-M,O,O, I mean to press and hold Alt, then type M, O, and O, in that order. Release Alt when you are done typing.

> These shortcuts have been tested on a 7290. Because of keyboard layout differences, some of these shortcuts will not work on a 7100 series device, such as the 7100g.

Navigation

Hold the Alt key while you use the trackwheel to scroll horizontally through any field where you can enter or view text. This works in many places: when you're editing a sentence in an email, a memo, or even a form on a web page. You can always move character by character by using Alt and scrolling the wheel. When viewing a message, list, or web page, you can scroll by page instead of by line by holding down the Alt key and scrolling. Hold the Shift key while you are scrolling through several items to select them all. This is useful for processing multiple messages at once. Table 1-1 lists some more navigation shortcuts.

Table 1-1. Navigation shortcuts

Shortcut	Description
Alt-Escape (the button on the side beneath the scroll wheel)	Changes the currently active application [Hack #6].
Alt-Enter	Locks the BlackBerry.
Spacebar or Return	Scrolls down a page at a time while reading an email.
Alt-Return or Shift-Space	Scrolls up a page at a time while reading an email.
B	Goes to the bottom of the message, list, or web page.
T	Goes to the top of the message, list, or web page.
N	Moves to the next item (next day in Calendar, next message in Messages).
P	Moves to the previous item (previous day in Calendar, previous message in Messages).
U	Moves to the next unread message.

Text Input

You may press Alt-Shift to activate Caps Lock. Of course, there is no need to use it in emails, since the BlackBerry capitalizes text for you automatically. When Caps Lock is on, an icon will appear in the top-right corner of your display that is a little oval with an arrow in it.

You may press Shift-Alt to activate number lock. This will let you punch the numbers on the BlackBerry keyboard with reckless abandon. The upper-right corner of your display will indicate number lock with a little oval icon with a "#" in it.

Home Screen

Table 1-2 lists the shortcuts that are available from the Home screen.

Table 1-2. Home screen shortcuts

Shortcut	Description
C	Compose a new message from the Home screen or when you're in the Messaging application. This will work with most firmware versions, but Vodafone maps "C" to be "Contacts" in theirs.
M	Launch Messages.
P	Launch Phone.
T	Launch Tasks.
L	Launch Calendar.
D	Launch Memopad.
F	Launch Profiles.
K	Lock the BlackBerry.
W	Launch the WAP Browser.
B	Launch the BlackBerry Browser.
E	Launch the BlackBerry Messenger.
A	Launch the Address Book or, on Vodafone devices, open the Applications folder.
R	Launch the Alarm program.
U	Launch the Calculator.
O	Launch the Options program.
S	Launch the Search program.
V	View Saved Messages.

Customize the Home screen. To rearrange the Home screen [Hack #5], highlight an icon, press Alt, and then click the trackwheel. You can move the icon somewhere else, hide it, or show all the hidden icons. This will let you hide "Enterprise Activation" if you don't need it, but it will not allow you to drop an item into a folder.

Sadly, there is no way to change or remap hotkeys to different applications.

View. Turn on the backlight by tapping the Power button. Some models, such as the 7290, have a two-stage backlight allowing you to set it to "off," "lit," and "well lit."

Messaging. You'll need some shortcuts in your Messages program once your emails start piling up. With your BlackBerry being so great at email [Hack #31], you'll love the shortcuts to slice and dice through your mail messages. The shortcut to select items is useful in selecting several messages at a pass. Your loving mother, who forwards you a few select jokes or chain letters every now and then, usually sends them in bulk. By using Alt-scroll you can roll your way down the list of emails from Mom and tap Del to delete them [Hack #22]. Sorry, Mom. You know I love you.

While reading an email you can press spacebar or Return to scroll down a page at a time, or Alt-Return or Shift-spacebar to scroll up one page at a time. Table 1-3 shows some more shortcuts you can use while reading an email.

Table 1-3. Email shortcuts

Shortcut	Description
Alt-I	Show only incoming messages.
Alt-O	Show only outgoing messages.
Alt-P	Show phone log.
Alt-S	Show SMS messages (or Alt-T in Vodafone firmware; presumably for "TEXT").
Alt-V	Show voicemail messages.
Spacebar	Page down.
B	Go to the bottom of the message.
T	Go to the top of the message.
U	Go to the next oldest unread message.
N	Go to the next message.
P	Go to the previous message.
Del	Delete selected or active message.
Alt-U	Toggle read status of selected message (unfortunately this doesn't work when you've selected multiple messages with Shift-scroll).
R	Reply to this message.
L	Reply to the sender and all recipients of this message.
F	Forward this message.
C	Compose a new message.

Table 1-3. Email shortcuts (continued)

Shortcut	Description
E	Go to next delivery error.
I	File selected message.
S	Go to search messages screen.
H	Scroll up until selected message is on the bottom of the screen.
J	Go to oldest message in the thread of the currently selected message.
K	Go to newest message in the thread of the currently selected message.
V	Go to saved messages.

While composing a message or PIM item. Press and hold a key to capitalize it. Press the spacebar twice to insert a period automatically and capitalize the next letter you type. This is one of the most underrated features of the BlackBerry.

Press the spacebar to insert @ and . while typing an email address; this works in fields that are known to require an email address.

Calendar Hotkeys

You can use hotkeys to change the view. You can, of course, pick the default view in the Calendar options, but occasionally you don't want to see things in agenda mode (which is my favorite). Table 1-4 lists the calendar shortcuts.

Table 1-4. Calendar shortcuts

Shortcut	Description
T	Today view.
D	Daily view.
A	Agenda view.
W	Weekly view.
M	Monthly View.
G	Go to a specific date.
C	Create new appointment.

Browser Hotkeys

The BlackBerry Browser has some surprisingly useful shortcut keys. Table 1-5 lists them.

Table 1-5. Browser shortcuts

Shortcut	Description
U	Toggle full screen mode.
I	View your history.

Table 1-5. Browser shortcuts (continued)

Shortcut	Description
S	Save current page to message list.
O	Access Browser options.
P	Show current page address and title.
A	Bookmark the current page.
D	Switch back to the program that previously had the focus.
F	Find text on current page.
G	Bring up the Go to dialog.
H	Go to your home page.
K	Bring up your bookmarks.
L	Bring up information about the currently selected link.
C	Bring up connection statistics about the current session.
Escape	Go back a page (like the Back button in your desktop browser).
N	Go forward a page.

Miscellaneous Keystrokes

You can change the signal strength meter from useless bars to the actual numeric of your signal strength. While in the Home screen, tap in Alt-N,M,L,L. For more information, see [Hack #17].

To get your BlackBerry to spill its guts and tattle on what version of software it has, battery level, PIN, IMEI, device uptime, and free storage, you can type in Alt-Shift-H (think "Help me!"). These are things you may need to have available if you call in for support due to a misbehaving BlackBerry.

To see the Event Log of your BlackBerry, type Alt-L,G,L,G. See "View the Event Log" [Hack #8] for more information.

—*Mark Rejhon, R. Emory Lundberg, and Dave Mabe*

HACK #4 Sync Your Calendar over the Air

Have your calendar items sync to your server-based mailbox continuously over the air.

If you are a BlackBerry Enterprise Server user, you know how quickly your emails arrive to your device—almost immediately. In earlier versions of BlackBerry Enterprise Server, synchronization of all items except email had to occur through a USB or serial connection using Intellisync with Desktop Manager. With the introduction of Version 2.1 of the BES, it is possible to update your calendar items over the air without cradling your BlackBerry.

Unless your device has been provisioned over the air to a 4.0 BES server, by default, your calendar items on your device are synchronized to your Outlook calendar only when you cradle and run an Intellisync using Desktop Manager. You can change that so your calendar items synchronize continuously over the air. This turns out to be quite a convenience for heavy calendar users and for users who have multiple people managing their schedule using a server-based calendar.

Enable Wireless Calendar

To set up wireless calendaring, you'll need to cradle your device and bring up Desktop Manager. Double-click on the Intellisync icon, and then click Configure PIM. Under Handheld Applications, select Calendar, and then click on the Choose button on the top-right side of the dialog box. As shown in Figure 1-5, instead of choosing Microsoft Outlook (which is the default), choose BlackBerry Wireless Sync, and then click OK.

 If BlackBerry Wireless Sync is not on the list, you either are not a BES user, your BES is using a version that doesn't support the wireless calendar feature, or your administrator has configured the server not to support it.

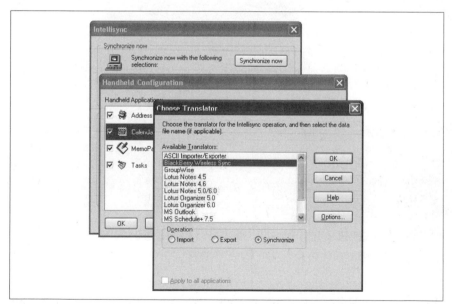

Figure 1-5. Configuring wireless calendaring

After clicking OK, a dialog box will appear saying that you need to synchronize to enable wireless calendaring. Because there are often many calendar items in your calendar that would take a lot of airtime and server resources to sync over the air, Desktop Manager has you do a cradled sync one last time before enabling the wireless calendar feature. For syncing a large number of items, this type of sync is far more efficient.

Click the Synchronize Now button to sync your device. When Desktop Manager recognizes that this is the first time you've synced since you first enabled the Wireless Calendar Sync, it will display a dialog box as shown in Figure 1-6. This simply explains that all your calendar items on your calendar will be overwritten with items from your server-based mailbox's calendar. Clicking OK starts the synchronization process as shown in Figure 1-7.

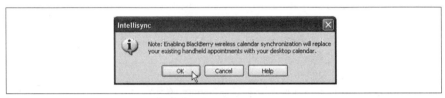

Figure 1-6. Warning about overwriting calendar items

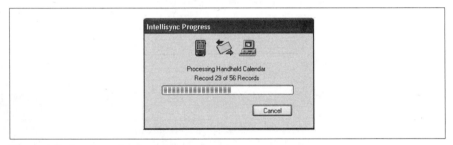

Figure 1-7. Syncing calendar items one last time

Once the Intellisync is complete, your calendar items will sync over the air whether the changes are initiated from your device or from your server-based mailbox.

HACK #5 Organize the Icons on the Home Screen

Remove your rarely used icons from your main screen.

Part of the success of the BlackBerry has been because of RIM's decision to use J2ME (Java 2 Micro Edition) as the OS on its devices. This decision immediately made a tremendous amount of J2ME software available for the BlackBerry and established the BlackBerry as a leading platform for application development. There are two chapters of this book dedicated to various third-party applications.

With all the programs that come with the BlackBerry software itself, plus all the third-party applications you're likely to install, it won't be long before you amass quite a collection. Each of these applications installs an icon on your Home screen consuming valuable screen real estate. These icons range from visually appealing to absurdly unrecognizable.

All these icons can make your Home screen difficult to navigate. Just when you get used to your icons being in a certain place, you install another application and all the icons shift around the screen so the Calculator icon shows up in the first instead of the last column. Chaos!

Fortunately, a little known feature of BlackBerry allows you to create some order in your icon wasteland by rearranging the icons and hiding others so they don't even appear at all.

Move Icons to and fro

Let's say you often search for messages from the Home screen by clicking the Search icon. You do it so often that you'd like to have that icon in a much more prominent position than the last row where it is by default. Here are the steps to move that Search icon to the first row right beside the Messages application.

1. From the Home screen, position the cursor on the Search icon (the magnifying glass).

2. While pressing the Alt key, click the trackwheel one time. A "hidden" menu appears as shown in Figure 1-8.

3. Select Move Application. A border appears around the Search icon, indicating you're moving the icon.

4. Use the trackwheel to position the icon beside the Messages application and press Enter. Figure 1-9 shows the Search icon in its new location.

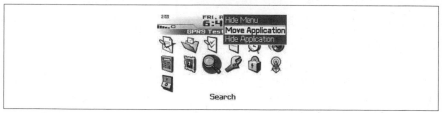

Figure 1-8. The "hidden" menu allowing you to move icons

Figure 1-9. The Search icon moved to its desired position

Completely Hide Icons

There are some icons I hardly ever use. They just take up space on my Home screen and get in the way. The minimalist in me likes to run a tight ship and show icons only for the applications I use most often.

To hide any icon on your Home screen, select the icon and repeat steps 1 and 2 from the previous section. Select Hide Application instead of Move Application. Figures 1-10 and 1-11 show how to hide the Memopad application.

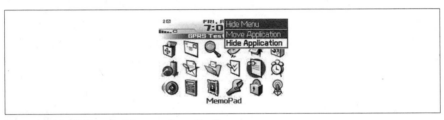

Figure 1-10. Hiding the Memopad application

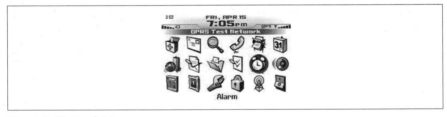

Figure 1-11. Poof! It's gone

Poof! The Memopad icon disappears, and the trailing icons shift over to take its place. This hack goes a long way to tidying up your Home screen.

Access a Hidden Application

"Uh-oh," I hear you say as you try this hack and realize that you'll actually need to access the Memopad application sometime. Don't panic!

BlackBerry makes it easy to unhide all your icons at once. Use the steps outlined earlier to access the menu you used to hide your icons. Once you've hidden at least one icon, the Show All option will appear on the menu (see Figure 1-12). Select this option and all your icons will now appear on the Home screen allowing you to access that hidden application. Notice each icon you've chosen to hide now has an X over it (see Figure 1-13).

Access the menu again to deselect the Show All option to hide the icons again.

Figure 1-12. The Show All option on the hidden menu

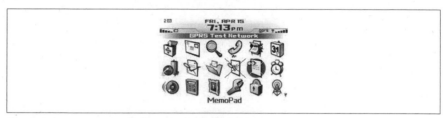

Figure 1-13. Accessing a hidden icon; notice the X

Run Programs in the Background

HACK #6

"Minimize" programs using the BlackBerry equivalent of Windows's Alt-Tab.

As BlackBerry users, we marvel at the efficiency that the BlackBerry provides. This efficiency causes even the slightest delay to seem like an eternity. Using the BlackBerry Browser to view a slowly responding web site can be frustrating when you're used to the quick and immediate access that the BlackBerry provides to most data.

It turns out that the BlackBerry OS provides an easy way to multitask by sending applications to the background. This method is analogous to the Alt-Tab functionality that Microsoft Windows provides.

Let's say you're using the BlackBerry Browser to access Painfullyslowsite.com. Rather than stare at a blank screen with a slowly moving progress bar across the bottom, you can respond to some emails and have the web site load in the background. To switch to the Messages application, hold the Alt key, and then press and release the Escape key. As you continue to keep the Alt key down, a list of icons appears, as shown in Figures 1-14 and 1-15. Each icon represents an application that can currently be switched to. Use the trackwheel to choose the program you'd like to switch to (the Messages program in this case), and then release the Alt key.

This sends the BlackBerry Browser to the background, and the Messages application is started and brought to the foreground. Similar to Microsoft Windows and other modern operating systems, these applications in the background continue to execute even though they aren't visible.

> This process is called *preemptive multitasking*, which means the operating system uses an algorithm to intelligently decide which application gets access to system resources such as processor and memory. Typically, foreground applications receive a higher priority for getting access to resources than a background application.

Figure 1-14. The Application Switcher on a 7290 device

Figure 1-15. The Application Switcher on a 7100 device

Convenient Sorting

The applications that appear on the Application Switcher list appear in an intelligent order, depending on which application you're running and whether there are programs you've already sent to the background. When no other applications are running, the Application Switcher shows the most common programs you'd likely want to switch to: the Home screen, the Phone application, and the Messages application.

The program you're currently in will appear first in the list, although the icon next to it will be initially selected.

As you send more and more applications to the background, they'll appear on the list as well. As you switch from program to program, you'll notice that BlackBerry conveniently places the icons in the order in which they were last accessed, with the most recently accessed application selected by default.

Because this functionality exists in the BlackBerry Operating System, all applications are able to use it—even applications made by third parties.

Switching More Quickly

When you're using one application at a time, the Escape key closes the current application and returns you to the Home screen. Once you have more than one application open, you'll notice that the Escape key still closes the current application normally, but it brings the most recently accessed program to the foreground instead of returning to the Home screen.

This serves a couple of different purposes. First, it is often exactly what you'd want to do. But second, it's also a nice reminder that there are other applications running that you've perhaps forgot were in the background.

This hack is probably most useful for minimizing the browser while accessing a slow site, but there are other convenient uses as well. The background programs are left in the same state they were in when you initially sent them to the background. This makes it especially useful for cutting and pasting text [Hack #2] between applications.

H A C K
#7   Type Less Using AutoText
Use abbreviations for commonly typed words to save time (and your thumbs).

You already realize by now that one of the main features that sets a Black-Berry apart from other handhelds is its keyboard. Its innovative keyboard design has won over millions of users because RIM was able to get a full QWERTY keyboard on a small handheld PDA. You can walk in any airport in the world and see that distinctive thumb-typing all around you.

But even as simple as the BlackBerry makes entering keystrokes, typing a long email is still cumbersome. A nifty feature called AutoText can alleviate some of that burden by replacing abbreviations with full text. This ends up saving valuable keystrokes by letting you type less.

Create a New AutoText Entry

RIM conveniently provides a slew of built-in AutoText entries that you may have already benefited from without even knowing it. For example, when you type "arent" into a new message, it immediately becomes "aren't." There are 108 built-in entries that provide corrections for common misspellings, add proper punctuation, and provide commonly used text snippets that you're likely to use when you type.

You'll definitely want to add your own custom entries to make shortcuts for text you commonly type. Suppose you often find yourself typing your mailing address into the mail messages you send via your BlackBerry. Here are the steps to create an AutoText entry to simplify that.

You'll find the AutoText feature in the Options section off the Home screen. Once you are in the AutoText program, click the trackwheel once to bring up the menu. Choose New (see Figure 1-16).

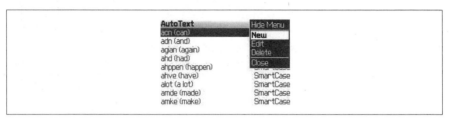

Figure 1-16. Adding a new AutoText entry

The configuration screen appears. Let's say you want your full mailing address to appear when you type "myaddr." Type "myaddr" in the Replace field, and then type your address in the With field, as shown in Figure 1-17.

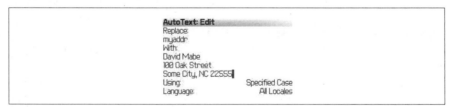

Figure 1-17. The new AutoText entry for your mailing address

The Using field has two options: SmartCase and Specified Case. SmartCase is BlackBerry's term for the feature that changes capitalization depending on the content you're typing in. This feature is a big time saver when you're typing normal text, but for your custom AutoText entries, you'll probably already know exactly the capitalization you want to appear, so you'll typically choose Specified Case.

The Language field is used to specify which language of the ones installed on your device this AutoText entry should work in.Click the trackwheel and choose Save.

> Using AutoText is great for entering commonly used phrases and text, but it's also great for common misspellings. If you find yourself often misspelling the same word, create an AutoText entry and just keep on typing.

Use Macros in AutoText

Not only can abbreviations be replaced with static text, but you can use AutoText to insert dynamic text as well. BlackBerry provides a set of 11 macros that can be used in the With field to insert the current date, owner information, or other variables. Table 1-6 shows a list of the built-in macros and what they do.

Table 1-6. AutoText macro strings

Macro string	Description	Example
%d	Short Date	4/13/2005
%D	Long Date	Wed, Apr 13, 2005
%t	Short Time	2:24p
%T	Long Time	2:24:35 PM
%o	Owner Name	As specified in device
%O	Owner Info	As specified in device
%p	Phone Number	18885551234 (from device)
%P	Device PIN	2100000a (from device)
%b	Backspace	(backspace)
%B	Delete	(delete)
%%	%	%

To access these macros, go to the With field in a new AutoText entry and click the trackwheel. Scroll down and select Insert Macro (see Figure 1-18). A menu will appear with the list of macros and a brief description.

Figure 1-18. The Insert Macro selection in a new AutoText entry

This hack is one you can use constantly. As you type, be on the lookout for text strings that would make good candidates for AutoText—your thumbs will thank you! An especially good use for AutoText is the text emoticons that pervade many of today's email messages. Those difficult-to-find symbols can slow your typing down to hunt-and-peck speeds. It's a lot easier to type "wk" than trying to type ;-) on a BlackBerry.

H A C K   View the Event Log
#8   Your device has a hidden Event Log that can be viewed using a certain key combination.

Similar to the event log on a Windows computer, there is an Event Log on your BlackBerry device where applications and the BlackBerry operating system itself can log information. Not only does this provide a central place to view all events from the system and applications, but it allows application developers to use an easy and consistent API for logging events so that each developer doesn't have to create his own.

You can view the Event Log on your device and even filter certain events, copy them to the clipboard—even email the entire log to someone. You won't find an icon for the Event Log viewer program. You'll have to enter a "secret" key combination to get it to appear.

View the Log

From your Home screen, type the following key sequence: Alt-L,G,L,G. This should bring up the Event Log viewer as shown in Figure 1-19.

```
Event Log (Warning)
W netrimudp - UCse
W netrimhrtRT - PNHr
a netrimhrtRT - ETot
a netrimotasync - DN+
a netrimhrtRT - PUpl
a netrimhrtRT - EPdp-0x0000000000
a ContentInjector - checking service bo
a netrimhrtRT - ESim-0x0000000000
a System - VM-RRnet_rim_app_mana,
```

Figure 1-19. The Event Log program

From within the Event Log, you can view the details of each event by pressing the Enter key. Figure 1-20 shows the details of an event.

You can copy specific events to the clipboard on your device by clicking the trackwheel when viewing the details of an event and selecting Copy Event. From the main viewer, you can copy a summary of the current day's events to the clipboard by using the trackwheel to access the menu and selecting Copy Today's Contents. Once the summary is on your clipboard, you can

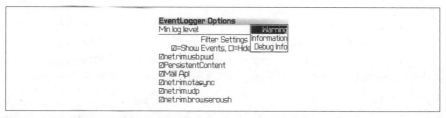

Figure 1-20. The details of a particular event

paste it into any other program, including in a new message, by using the trackwheel menu and choosing Paste.

> When you access the Event Log, the program reads the current events and displays them on the screen. As you are viewing the events, additional events may have been logged since you started the program. Choose Refresh from the trackwheel menu to reread the events to make sure you're displaying the most recent events.

Customize Event Log Options

You can use a number of options to filter the events or even expand your view to include events that have a lower severity. To access the filter options, click on the trackwheel and choose Options from the menu. By default, the Event Log displays events with a severity of Warning. You can change the Minimum Log Level setting in your Event Log as shown in Figure 1-21.

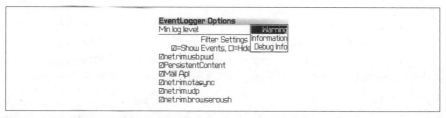

Figure 1-21. Changing the minimum log level

> Modifying the Minimum Log Level changes the threshold for any events that get logged after you make the change. So you cannot change the level from Warning to Information or Debug and expect to see additional events retroactively—only new events will be affected by your change. Also, consider the impact of setting the threshold to Debug Info: as with debugging logs on any platform, they can quickly fill and consume system resources.

You can also control which applications' events show up in the Event Log. By default, the events from all applications are displayed in the Event Log. To modify which program events appear, uncheck the checkbox beside the applications you'd like to not appear in the log. Be sure and save your changes after you make modifications to the Options screen. This is useful for troubleshooting a specific application you're having problems with.

The Event Log program also gives you the ability to clear the log. Choose Clear Log from the trackwheel menu to purge all entries in the Event Log as shown in Figure 1-22 to start from scratch when troubleshooting a problem.

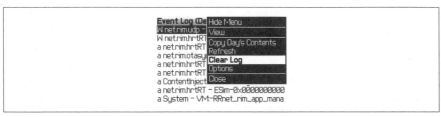

Figure 1-22. Clear the Event Log

Use Your BlackBerry as a Modem

HACK #9

When you're in a bind, tether your laptop to your BlackBerry and connect at modem speeds.

Imagine you are in a remote location, far, far away from the nearest WiFi hotspot or Ethernet connection. You have a deadline and you have to send a proposal from your laptop as soon as possible. Sure, you could print out the proposal and fax it, but where is the fun in that? The BlackBerry 7290 and 7100 come with a built-in modem that can be accessed through the USB cable. It won't connect at EDGE or EV-DO speeds (yet!), but this hack can come in handy in certain situations.

The technique of using a mobile phone's data connection [Hack #1] is commonly known as "tethering." Normally this is done through a wireless Bluetooth connection on a mobile phone. Despite the Bluetooth capability on recent models, the modem on the BlackBerry can, unfortunately, only be accessed through the USB cable.

Set Up a Dial-up Connection

You'll need to set up a dial-up connection on your Windows machine to use the BlackBerry modem. When you installed the BlackBerry Desktop Manager, it should have automatically installed the modem drivers to use with your device, and you should have a modem called "Standard Modem" in the

Phone and Modem Options section of your Control Panel. The drivers are located in the following folder: *C:\Program Files\Common Files\Research In Motion\Modem Drivers*. Use the following steps to set up your connection:

1. Connect your BlackBerry to your computer using the standard USB cable that came with your device and start the BlackBerry Desktop Manager software.

2. Go to the Control Panel, then Network Connections, and select Create a New Connection.

3. Select Connect to the Internet, then Set up my connection manually, then Connect using a dial-up modem, and select your BlackBerry modem from the device list.

4. Choose a descriptive name for the ISP Name (e.g., BlackBerry Modem Connection).

5. Enter the phone number for your service provider. You'll probably have to check with your carrier to get the right number. For Cingular and other GSM carriers (AT&T, T-Mobile), it's *99#. For CDMA providers (Sprint, Verizon), the number will be #777.

6. The username and password will vary by carrier as well. Cingular just accepts null values for both fields.

7. Complete the wizard, being sure to deselect the "Make this the default connection" checkbox.

Set Up Your BlackBerry Modem

If you have CDMA service, you can skip this step. If you have GSM service for your BlackBerry, you will need to add an extra init string for the modem to use when it connects. This string is specific to each provider, so you'll have to contact your provider for the proper setting. Go to the Modems tab in Phone and Modem Options in Control Panel. Find the "Standard Modem" that was installed with your BlackBerry Desktop Manager installation and double-click it.

On the Advanced tab (see Figure 1-23), you'll need to enter an extra initialization command that is specific to your carrier. Enter +cgdcont=1,"IP","APN" where *APN* is your carrier's access point name.

> The APN, or Access Point Name, is the name of the network to which a GPRS device connects. Check with your provider to get the value that should be used for your service. There is also a fairly comprehensive list of APNs by carrier at the following URL: *http://www.opera.com/products/mobile/docs/connect/*.

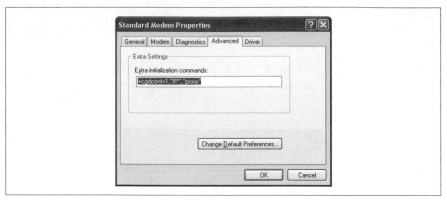

Figure 1-23. Entering the extra init commands with the APN specific to your provider

Establish the Connection

You should be able to use the new dial-up connection in Network Connections to establish an Internet connection. You can right-click on your new connection and choose Connect. Click Dial without entering a username or password. If all goes well, you will have an active wireless Internet connection!

The data connection that you get from your carrier typically provides access for the most popular Internet protocols (HTTP, HTTPS, POP3, IMAP, SMTP, etc.). Access to anything beyond those major protocols may or may not be allowed. Your mileage will vary widely.

Connection Speeds

Figure 1-24 shows a recent connection I've made on a Windows XP machine.

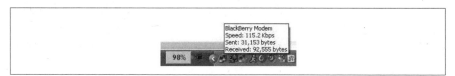

Figure 1-24. Connection statistics for your BlackBerry modem

You'll notice that Windows shows that I connected at 115 Kbps—that's pretty fast for a GPRS connection, right? Actually, the value you'll see for the connection speed is a little misleading. It's referring to the connection speed between your laptop and the phone. The actual throughput you'll experience is capped at 40 Kbps for GSM networks and roughly 143 Kbps for CDMA networks. To get a reasonable estimate of the realized bandwidth on your GPRS connection, you can use the Mobile Speed Test, available at *http://text.dslreports.com/mspeed*.

Pay attention to the details of your data plan, especially when using this hack. As you can imagine, you can consume a good portion of your monthly data allotment using your BlackBerry as a modem. Of course, if you have an unlimited data plan, there's really nothing to worry about (unless you are paying for roaming).

HACK #10 Maximize Your Battery Life

Use these settings to disable the battery-intensive functions—your battery will thank you.

As far as batteries on mobile devices go, the BlackBerry's is pretty good. But we all realize that the innovation in the battery industry seems to have almost stagnated. We have seen hard drive capacities skyrocket while their prices have plummeted. The same cannot be said for the capacity of the batteries we use.

We, as BlackBerry users, expect a lot from the batteries in our devices. We want to be able to go on a trip and not worry about remembering the travel charger. And heaven forbid we have to turn wireless off on our device to conserve the battery—oh the horrors!

This hack explains which functions should be avoided when you know you're going to be without a charger for a while and you need to stretch your battery life to the limit.

Turn wireless off
> I know we hate to admit it, but the wireless modem in your device is far and away the worst battery hog of all. This will probably be called blasphemous by some, but when you're in a bind, turn the wireless off. In my secondary device, I once turned the wireless off with a full charge and left the device on. When I checked it again a full two weeks later, the device still showed full strength.

Never use the backlight
> The backlight consumes a good deal of the battery, too. It sure is easier to look at the screen with it on, but avoid using it when you are in situations when you need your battery to last. Of course, you don't have much choice if your device is one of the 7100 models that only have a backlit screen.

Keep your device in the holster
> When your device detects that it is in its holster (through clever use of a magnet), it immediately goes into a power save mode. The backlight timeout is ignored, the screen is turned off, and it operates on minimal power while in the holster.

Don't keep programs running in the background
> Any programs that are running in the background on your device [Hack #6] will require and use system resources to run. Keep these to a minimum.

Avoid programs that require wireless use
> Obviously, it's not always possible to completely shut off the wireless modem in your device. When you have to have it on, minimize its use. That means avoid using programs that use it. This includes the Black-Berry Browser and many of the third-party programs outlined in this book. Instant messaging applications are notorious for frequent use of the wireless modem while running in the background. This is a serious drain on your battery!

Keep plenty of free space available
> The BlackBerry operating system manages memory similar to a PC. Memory is swapped to a flash memory disk as RAM becomes scarcer. This swapping can be resource intensive and therefore can eat away at your battery's charge. Go to the Status item in Options and look for the File Free field. RIM recommends maintaining a minimum of between 400 and 500 KB free for optimal usage [Hack #12].

Turn Bluetooth off when not in use
> Just like the wireless modem, the Bluetooth function will consume a tre-mendous amount of your battery. Keep it off unless it's in use.

Turn the device off completely
> This is the best way to conserve the battery! Use the Auto On/Off func-tion in Options to have the device automatically shut off and start back up on a schedule you specify. You can even set up a different schedule for weekdays and weekends.

HACK #11 Optimize Your BlackBerry Browser
The browser defaults are certainly not optimal. By changing the settings on your handheld, browsing can be much quicker.

The relatively recent addition of the BlackBerry Browser to the device makes it easy to browse web sites wherever you happen to be. Some might argue the built-in browser is probably the most underused program on the device.

By default, the browser will download all images, JavaScript, and stylesheets for each web site you visit. Although some sites built with XHTML look quite nice on a handheld, there are many sites that haven't gotten on the standards-compliant bandwagon. Many of these sites look plain silly on a handheld or are even downright unusable.

You can customize the settings on your BlackBerry Browser to provide a quicker, smoother ride no matter what virtual terrain comes your way.

How the Browser Handles Images

Many of the images in the web sites you visit aren't optimized for handheld viewing. For each image on a web page, your device will have to initiate another HTTP request to retrieve it. On pages with just a few images, this is a serious time consumer.

When you visit a web page with images, the text of the page is rendered first and then the images are retrieved and rendered in your browser. When your browser initiates the request for a new image, it has no way of knowing the size it should allocate on the screen for that image, so it ends up allocating a very small amount of screen real estate for it.

When the image is downloaded to your device, the browser renders it and displays it on your screen. This can wreak some serious havoc when you're viewing a web site. Images take much longer to load than the text on the page; therefore, it's quite possible you've already read a good portion of the text on the page and then—bam! The image is loaded and the paragraph you were right in the middle of "disappears," because the browser has allocated the proper amount of space for the image, displacing the text. How frustrating! You now have to scroll down and find where you were reading when you were so rudely interrupted. This often happens more than once on a page that has multiple images!

Use "Minimal" Settings on the Browser

You can turn off several features of the browser that cause it to make separate HTTP requests that take up valuable time and battery power. Go to the browser and click the trackwheel once to bring up the menu. Choose Options.

The options we are looking for can appear in a couple different places, depending on which version of the browser you have loaded. They will be on either the Browser Configuration section or the General Properties section, or a combination of the two. Look for the Show Images option as shown in Figure 1-25.

Figure 1-25. Image settings in the browser

There are three options for the Show Images setting:

No
> Don't retrieve or show images for any type of page.

On WML Pages Only
> Show images only for pages in Wireless Markup Language format.

On WML & HTML Pages
> Show images in basically all web sites.

For optimal settings, avoid the default of On WML & HTML Pages. I choose to load no images for any type of page. The On WML Pages Only is also a reasonable choice. WML pages are formatted specifically for handheld devices, so the images are already lean and mean and optimized for your device.

The other settings that will streamline your browsing pleasure are shown in Figure 1-26.

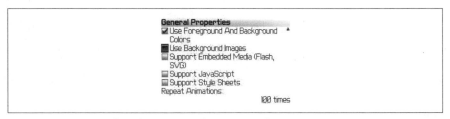

Figure 1-26. Settings for other resource hogs in your browser

Be sure and uncheck the checkboxes to disable the following settings:

- Use Background Images
- Support Embedded Media (Flash, SVG)
- Support JavaScript
- Support Style Sheets

Each of these types of formats causes your browser to initiate a costly, relatively resource-intensive HTTP request for fancy content that, the majority of the time, isn't worth the effort.

Enable Image Placeholders to Retrieve Images Only on Your Schedule

In Figure 1-25, notice the Show Image Placeholders setting. By default, this setting will be turned off. Turning it on provides a valuable function that will come in quite handy as you "pimp your browser."

When you've configured your browser to show image placeholders, you'll see a small icon where each image would go were you to download it. If there is alternate text for text mode browsers, you'll see that beside the icon; otherwise, it will simply say "image." If you come across an image that you actually would like to download, there are two menu items, shown in Figure 1-27, that allow you to override your browser settings and download images on the current page only.

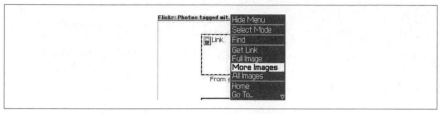

Figure 1-27. The additional options to selectively download images

As you are viewing a web page with an image you would actually like to download, click the trackwheel once to bring up the menu. Use the More Images option to download the current image that your cursor is positioned on (see Figure 1-28). Alternatively, you can use the All Images item to retrieve all images in the current page. This lets you have the best of both worlds—the lean speed of a text-only browser with the ability to view an image on demand.

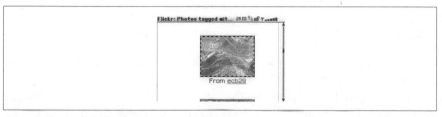

Figure 1-28. A downloaded image in the browser

HACK #12 Maximize Your Free Memory

Use these settings so you'll always have enough memory for the latest third-party application you want to install.

As you receive more email, install more applications, and save more pictures and ringtones to your BlackBerry, you'll continue to consume more and more RAM on your device. Some of the devices RIM released in the early days had just 2 MB of storage or even less. Those of us who had one remember having to hand-crank them and trudge through the snow, uphill both ways to get them to work. Maybe I exaggerated a little, but they've come a long way.

Maximize Your Free Memory

Unfortunately, there is still only a finite amount of RAM available on devices and a seemingly unlimited supply of third-party software for the BlackBerry. This will require that you manage your storage space on your device.

Not only will you need to keep a comfortable amount of disk space available to install additional software, but there are important performance reasons to do so as well. The BlackBerry operating system uses the lightning-fast *SRAM* for normal memory operations. When that memory runs out, it uses the much slower *flash memory* for swap space. This is analogous to your PC, which will use a swap file on your computer's hard disk when it runs out of hardware memory. Just like your PC, your BlackBerry will slow down considerably if the amount of available swap space is too small.

Determine How Much Free Space You Have

To see how much free space you have available, go to Options → Status. This shows you the total amount of storage space you have in your device and the amount of free space, as seen in Figure 1-29.

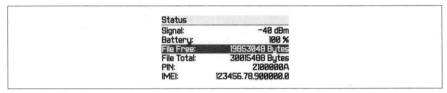

Figure 1-29. The amount of total space and free space available

RIM recommends having at minimum of between 400 and 500 KB (400,000 to 500,000 bytes) of free space available for the operating system to swap memory efficiently. Any less than that and you'll stare at an hourglass for an excessive amount of time for whatever function you perform.

What Data Uses the Most Space?

So you have gone to the Status area and realize that it's time for spring cleaning. How do you determine what needs to be cleaned up? What can you delete to get the most space quickly? And we need to know fast because there's a game of Texas Hold 'Em [Hack #32] and its time to ante up!

Although it takes a little work to get there, there is a way to view the amount of storage space that each application is consuming on your device. You'll need to complete these steps from your PC with your device in the cradle.

Bring up Desktop Manager and double-click on the Backup and Restore option. Click on the Advanced button to bring up the advanced backup and restore tool. On the Handheld Databases section of the tool, select all the databases in the window. Click the arrow pointing to the left to create a backup of all those databases. This will take some time to complete.

Although the main function of this tool allows you to selectively back up your handheld databases, as a side benefit, you can use the Bytes column once the backup is complete to determine which programs are using the most space. See Figure 1-30.

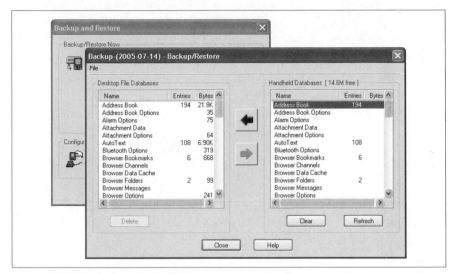

Figure 1-30. Using the advanced backup restore tool to determine storage consumption

How to Delete an Application

If you determine that you no longer need an application, go ahead and delete it. Go to Options → Applications on your device and view the installed applications. Click the trackwheel once to bring up the menu and choose the Delete option, shown in Figure 1-31. This will remove the application from your handheld. In most cases, you will need to reboot the device to actually purge the application and free the space it was using.

Figure 1-31. Deleting an application

Maintain Your Free Space

There are a couple of settings to control the two heavy hitters on your device: your messages and your calendar items. By default, the messages on your device stay around for 30 days before they are deleted automatically. You can control this setting by going to your message list and clicking the trackwheel once. Choose Options from the menu that appears and click on General Options. You can use the Keep Messages option, as shown in Figure 1-32 to choose whether messages are kept on the device for 15, 30, 60, or 90 days. You can choose to keep your messages on your device forever although, obviously, that is not recommended.

General Options	
Display Time:	Yes
Display Name:	Yes
Confirm Delete:	Yes
Hide Filed Messages:	No
Make PIN Messages Level 1:	Yes
Auto More:	Yes
Keep Messages:	30 Days

Figure 1-32. Control how long messages stay on your device

In your Calendar application, there is a similar setting called Keep Appointments where you can choose to delete appointments after a set amount of time has elapsed since the appointment.

 If you're a BlackBerry Enterprise Server user and you're using wireless reconciliation and/or the wireless calendar **[Hack #4]**, you may be wondering what happens to those messages when they "expire" off your handheld. Fortunately, the deletions that occur as a result of this setting are not reconciled with the server, so you'll be able to manage the data on your server mailbox separately.

HACK #13   Get Things Done with the BlackBerry
Kick butt and take @names.

The David Allen Company has a fantastic productivity system called Getting Things Done, also known as *GTD* (*http://www.davidco.com/what_is_gtd.php*).

I will say this: Getting Things Done is a remarkable way to restore some balance and control over your life, and not just in the office. It works great for all occasions, and has managed to bring my stress levels from catastrophic highs to a much more reasonable pitch. It is easily the best system that I know of for *chaos reduction*.

Everyone implements GTD a little differently. I don't think I've ever met a purist in this dark art, with the full set of tickler files, folders, and all the other paraphernalia. It is a system that can be tooled up to work pretty well with the BlackBerry, however, and it doesn't even have to be painful.

This is a system that I cannot resist fiddling with a little bit, and most GTD nerds with an engineering and tinkering background do the same thing. These days I'm using a BlackBerry 7290 on T-Mobile, BES and BIS, and Microsoft Entourage. I am also using Microsoft Outlook 2003.

Set Up Your Categories

In both Outlook and Entourage, I have created categories for various contexts that I use in GTD. This includes but is not limited to:

@home
> Things to do when I'm at home or in the home office.

@office
> Things to do when I'm at the work office. The one with the bad lighting and a parking shortage.

@calls
> Things to do when I'm able to make phone calls but nothing else, or just calls I need to make.

@anywhere
> Things I can do anywhere, any time. Example: Think about ideas for *BlackBerry Hacks*.

@errands
> Things that I need to do when I'm already out and about. Example: Buy Silk Chocolate Soymilk at Whole Foods.

 Geeks tend to prefix their GTD categories with letters that cause them to sort first in file listings. Hence, the @ in all the names. It lets you sift out the GTD stuff from all the other detritus in your PIM and on your filesystem.

Organize Your Notes

In the Notes application on the BlackBerry, I also have a few lists of things I want to talk to people about. I flag those agenda items by titling them as *#Name*, so that #BBKing is a note that has things I want to talk to BB King about should I run into him in the hall.

Decide When to Create a Task

If it is something more pressing, that becomes a task. If it has to be done on a certain day, it becomes an all-day event on the Calendar. If it has to be done at a certain time on a certain day, it becomes a traditional appointment with a real date, time, and reminder.

The BlackBerry does a good job at facilitating this as long as you're using BlackBerry 3.8 or 4.0, since these versions support categories.

Since I have created all those @contexts in my PIM, I must use them on every single individual task I have in my Big List of Tasks. If you don't do this, you can easily lose track of what you should be doing!

Search and Sort

I also have a lot of projects assigned in the Entourage Project Center. It watches folders, links items to relevant data, and otherwise gets out of the way from me trying to get things done. What PocketMac for BlackBerry [Hack #61] does here is pretty neat: it calls projects in Entourage *categories* and syncs those along, too. This means that I can surf a Task list and filter based on category as expected, but in GTD land, this lets you filter by project.

If I'm at home and I want to know what I have waiting for my attention, I just filter my Tasks to show me everything categorized @home. If I want to work on a specific project, I can also filter on the category BlackBerry Hacks, and see all the waiting tasks assigned to that project regardless of context (see Figure 1-33).

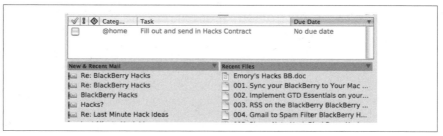

Figure 1-33. Viewing a project and viewing a task for @home

Some people will tell you that you should include the context in the title of your task, and not create categories for each context. That sounds pretty good in theory, but if you have multiple contexts for a given item, you're going to have items that are called "@home & @calls Call Insurance Company about Flood Damage to Basement" and the BlackBerry can only show you a portion of this information. I choose to maximize my visibility by cutting out the prefix of @context and using categories for each context and the filter to show me what is relevant to my current state.

Don't Interrupt Yourself

It turns out I don't respond well to nagging interruptions from a PIM or handheld device. Go figure. GTD breaks things out into tasks and projects so that you can do micro-tasks on the path to a final outcome of your choosing, which is really nice. It also forces you to work things in smaller chunks, which is even nicer. For example: instead of a task that says "Write hacks for BlackBerry Hacks," I go ahead and scribble that down in a notebook. I think and scribble a little bit about what that means, and those items become individual tasks. I may want to brainstorm about what I think should be in a Hacks book about the BlackBerry. I may have already written a bunch of information already that I can gladly hand over to the editor and author of the book. All of those steps become tasks under one project, and as I do them, I tick them off.

You see? Instead of staring at "Write hacks for BlackBerry Hacks" for a month, I actually had a long list of items that get me there. The tasks were meaningful and relevant and some of them could be done at home, some in line at the supermarket, and others required an errand. The BlackBerry presents me with options when I pull it out, based on things I can actually do depending on where I am. That is quite possibly the most important thing to remember about GTD: You get to pick what you want to do any time you want to do something, and it is always relevant. No more noise!

I don't use the typical GTD "waiting-for" lists because, in most situations, I can note what I'm waiting for in the task detail instead of having a separate list. Your mileage may vary on this, and you may very well decide to keep a waiting-for list in your Memopad.

Everyone uses GTD a little bit differently. It is my hope that this hack will help jump-start your newfound life as a productive and contributing member of society.

For more information and tips on applying GTD to a geek's life, see Merlin Mann's 43 Folders weblog at *http://www.43folders.com/*.

—*R. Emory Lundberg*

HACK #14 Create Your Own Polyphonic Ringtones

With a little work, you can create your own polyphonic ringtones for your 7100 series device.

Who would have thought that ringtones would have become the industry that it is today? Ringtones are a $270 million industry, and that number is expected to balloon to $724 million by 2009 according to Jupitermedia. Everywhere you go you hear another mobile phone user's silly ringtone.

Audio has had mostly primitive support on the BlackBerry—that is, until the introduction of the 7100 series that is targeted toward the consumer as opposed to the business user. The 7100 series devices have support for polyphonic ringtones, and the 7290 device has support for monophonic ringtones. This hack will show you how to get your own custom polyphonic ringtone on your 7100.

There are five steps you'll need to take to get an audio clip in the right format and onto your device.

1. Get your audio into WAV format, either by recording your own sound using your computer and a microphone, ripping a track from a CD, or converting an MP3 file to WAV format.

2. Use a tool such as Audacity (*http://audacity.sourceforge.net/*) to crop your audio clip to a reasonable time for a ringtone, probably about 10 seconds.

3. Convert your audio clip to the Oki ADPCM audio format.

4. Put the converted audio clip on a web server that contains the appropriate file type for ADP files.

5. Use your BlackBerry Browser to access the URL of the file you placed on the web server.

If this sounds like too much work for you (not everyone has access to a properly configured web server), there is a service that you can pay for that will convert your audio clips and host them on a web server for you to download to your device. The Vodaberry service (*http://www.blackberrytunes.com*) costs just $18 a year for an unlimited number of MP3 conversions.

Format Your Audio

You'll need to somehow get your audio in WAV format. There is a variety of ways to do this and, in some cases, no formatting will be needed. This format is fairly standard—it is the default format of the Windows Sound Recorder program, and most any audio program will support it. Just to get your feet wet, create a short audio clip with the microphone on your computer.

You can use Audacity to select about 10 to 15 seconds of audio and save that clip to a file. Figure 1-34 shows how I selected 15 seconds of audio from my perfectly legal copy of "Free Bird" by Lynyrd Skynyrd and used the Trim feature of Audacity to shorten the clip.

Encode into Oki ADPCM Format

Once you've shortened your audio to a length suitable for a ringtone, you're ready to convert the clip to a format that your BlackBerry can use. While this seems like it should be a reasonable task, there is only one software

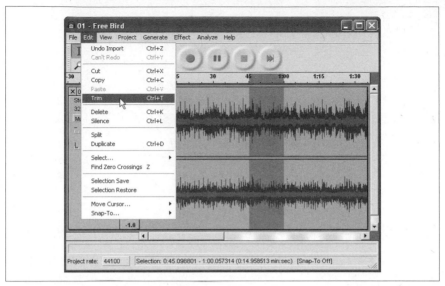

Figure 1-34. Trimming audio in Audacity

package that is able to convert audio into the Oki ADPCM format that the 7100 series devices use for polyphonic ringtones. It is called Mobile Phone-Tools and is currently available for $39.90 at *http://www.bvrp.com/eng/products/mobilephonetools/*. This software package is intended for use with specific mobile phones, communicating with them over a USB cable. You will need to specify a mobile phone that is supported by the software to purchase it. At the time of this writing, BlackBerry phones were unsupported, so pick something at random (or another phone that you own). For example, selecting a Phillips 530 will get you to the purchase page just fine.

After downloading and installing Mobile PhoneTools on your Windows-based computer, run the program from the Start menu. Since it is made to synchronize data between your computer and non-BlackBerry mobile phones, it will ask you to configure a connection to your phone. Click Cancel to skip this step and go straight into the program. Click on the Menu button and select Tools → Multimedia Center as shown in Figure 1-35.

Click on the Melody Studio tab on the left side of the program. In the center section of the screen, browse your local filesystem to find your audio clip and double-click it. Click the Save As button and choose the *Oki ADPCM* file type in the Save As Type drop-down, as shown in Figure 1-36.

Get the Ringtone on Your Device

So you've got the ringtone in the right format, but how do you get it onto the BlackBerry? The only way to get a ringtone on your device is by using

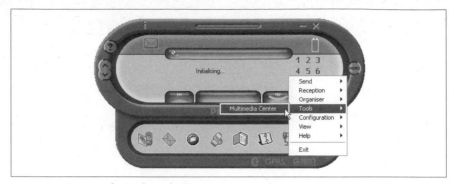

Figure 1-35. Go to the Multimedia Center

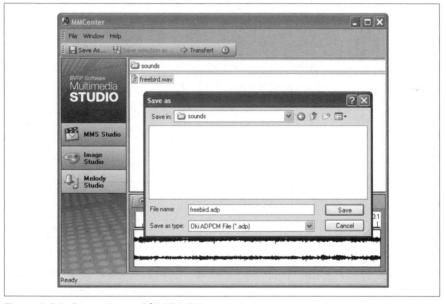

Figure 1-36. Converting to Oki ADPCM

the BlackBerry Browser, so you'll have to get the file onto a web server and access it through a URL. The BlackBerry Browser expects to see a MIME type of *audio/adpcm* for polyphonic ringtones; otherwise, your ringtone will be treated as plain text, which won't work.

> This is not a default MIME type on either Microsoft's IIS server or Apache, so you will probably need to add the MIME type and restart the web server. Just add the audio/adpcm MIME type for files with a *.ADP* extension.

When you access the URL for your ringtone, the BlackBerry Browser will download the entire ringtone and automatically play it. There is a 128 KB limit for ringtones—if you try to download one bigger than that, you'll get an error message.

Depending on your wireless connection, this can take several seconds to download. Just when you least suspect it, your device will start playing the ringtone, so don't do this in a library or other place where you'll cause a scene (unless that is your goal).

You'll need to save the ringtone to be able to choose it for your ring in your Profiles. Click the trackwheel and choose Save from the menu. Choose a descriptive name for the ringtone or just choose the default filename and select OK, as shown in Figure 1-37.

Figure 1-37. Saving your ringtone

After saving it, you will be able to choose your newly crafted ringtone from the Tune option in your Profiles (see Figure 1-38).

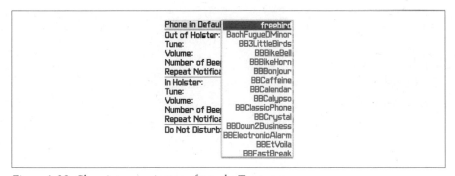

Figure 1-38. Choosing your ringtone from the Tune menu

You can even choose to play "Free Bird" whenever you receive an email message on your device! (Oh, the irony.) I've provided a sample ring tone of a rooster crowing at the following URL: *http://dave.runningland.com/rooster.adp*.

Store a Photo Collection on Your Device

#15   Save images to your device and choose ones to display in prominent positions.

Beginning with BlackBerry 4.0, the handheld software ships with an application called Pictures. This program comes with a few stock photos to get you started. You can also add your own images to your device, although it is not at all obvious at first what steps you need to take to get them there. Use this hack to add your own pictures to your device and add them to your background.

Use the Picture Application

To use the Pictures program, from your Home screen, click on the Pictures icon to bring up the program, as shown in Figure 1-39. The application displays a list of the images that you have on your device. From this screen, you can choose a picture and do several things with it via the menu.

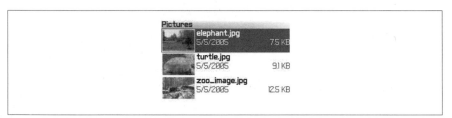

Figure 1-39. The Pictures application

Use the trackwheel to bring up the menu. Here are the options available on the menu and their functions:

Set As Home Screen Image
Use this option to have the current picture display as a background on your Home screen.

Reset Home Screen Image
Reverts your Home screen background image to the factory default.

Set As Standby Screen
Display the selected picture as the image on your standby screen. This is the screen that displays your owner information when your device is locked. The 7100 series devices don't have this option.

Reset Standby Screen
Reverts the standby image back to the factory defaults. The 7100 series devices don't have this option.

Open

> Open the selected image in a full-screen view.

Delete

> Delete the selected image from the device, freeing up valuable storage space.

See Figures 1-40 and 1-41 for examples of using an image for a Home screen background and standby screen background. Note that when using a background image for the Home screen, the BlackBerry will fade the image slightly to accentuate the icons which, of course, is the main reason why you use the Home screen anyway.

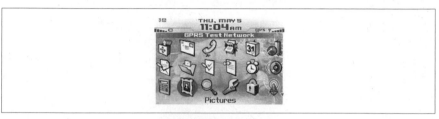

Figure 1-40. A custom Home screen background image

Figure 1-41. A custom standby screen image

Put Your Own Images on Your Device

So how do you get your own images on your device and use them for your backgrounds? It's not as easy as it probably should be. You can't attach a picture to an email and send it to yourself; you'll have to use Berry Pix [Hack #69] to do that. The only way to add custom images to the device is through the BlackBerry Browser. You have to actually browse to a web site that has the photo that you'd like to save on your device.

Once you find a picture on a web site that has an image you'd like to store on your device, you need to select the image. Use the trackwheel to scroll to the image until it is selected. You'll know it is selected when a dotted line surrounds the image. Click the trackwheel once to open the menu and choose Save Image as shown in Figure 1-42.

Figure 1-42. When an image is selected, a Save Image option appears on the menu

But how do you get your own images on the Web? There are several ways: use your web space (if you have one), your weblog, or a photo-sharing service.

Many Internet service providers include web space in their packages. Upload your pictures to your space, and visit the site from your BlackBerry. You could even start a blog and upload pictures as you post blog entries.

There is also a popular, innovative service called Flickr (*http://www.flickr.com*) that is perfect for this purpose. Flickr allows you to upload images to the Web and then tag them with keywords that describe the photo. For example, I could upload the turtle picture and tag it with "zoo" and "turtle." You can then search for or subscribe to photos that anyone has tagged with certain keywords. Because Flickr promotes an online community, it has quickly become quite successful.

See Also

- Gallery (*http://gallery.menalto.com*) provides open source software for displaying photos on a PHP-enabled web site.
- GIMP (*http://www.gimp.org*) is cross-platform, open source software for editing images.

HACK
#16

Turn Your Computer into a Speakerphone

If you have a BlackBerry that supports Bluetooth, use your computer as a poor-man's headset.

Up until fairly recently, RIM didn't make a BlackBerry device that supported Bluetooth, even as it is becoming a standard feature of other smart phones. That all changed when RIM released the 7290. The BlackBerry 7290 as well as the 7100 series devices have built-in Bluetooth, and you can expect future devices to include support for the protocol as well.

Given all the data synchronization capabilities of the BlackBerry, you'd expect to be able to do lots of things with the Bluetooth adapter on the devices—but as of now, the capabilities are rather limited. You can't tether a laptop to share its data connection using Bluetooth—you have to cradle the device for that [Hack #9]. However, you can use a Bluetooth headset so you don't have to deal with the cord on your earpiece—or worse, hold that oddly shaped 7290 to your ear to talk on the phone. The fancy Bluetooth headsets are plentiful but still somewhat costly. If you've purchased a computer recently, there's a decent chance it already has Bluetooth built in. If not, you can purchase a Bluetooth adapter for a reasonable price that allows you to do a whole host of things with computer peripherals, including acting as a headset for your BlackBerry!

> Some older Bluetooth adapters made for PCs have limited Bluetooth support. Be sure that the adapter you are buying supports the Bluetooth Headset Profile. Most, if not all, Bluetooth 1.1–compliant adapters should do the trick.

Wait, you say, isn't the goal to be *more* mobile—why would I want to use my computer as a speakerphone? And here comes the standard hacker's response: "Because you can!"

Install a Bluetooth Adapter

You'll need a Bluetooth adapter with the popular WIDCOMM chipset and software. Other chipsets may or may not work. I bought a Kensington Bluetooth USB Adapter from my local electronics store that came with the WIDCOMM software. You'll need to install the software on your Windows XP computer before inserting the USB adapter.

After installing the adapter, you'll find a new icon on your desktop called "My Bluetooth Places" as well as a Bluetooth logo on your task bar, as shown in Figure 1-43.

Figure 1-43. The Bluetooth icon on the task bar

Pair with Your BlackBerry

By default, the Bluetooth adapter on your BlackBerry is disabled. You'll need to enable and set its status to "discoverable" so that other Bluetooth adapters within range will be able to see it. On your device, go to the Options program and go to Bluetooth. Use the trackwheel to bring up the menu and choose Enable Bluetooth. After a few seconds, your device will display a dialog saying it has been enabled, and you'll return to the Bluetooth screen with a list of Paired Devices. Use the trackwheel to select Options from the menu. You'll need to change the Discoverable status to

Yes. Optionally, you can change the Device Name that your device will advertise itself as. Click Save from the trackwheel menu.

Once you've enabled Bluetooth on your device, go back to your desktop computer and right-click on the Bluetooth task bar icon and go to Bluetooth Setup Wizard. This brings up a wizard (see Figure 1-44) that will assist in discovering Bluetooth-enabled devices and configuring them.

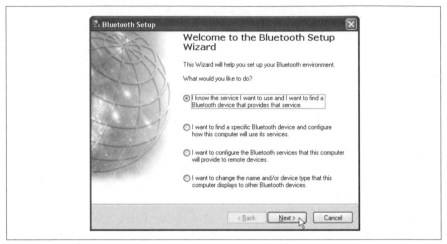

Figure 1-44. The Bluetooth Setup Wizard

Choose the "I know the service I want to use and I want to find a Bluetooth device that provides that service" option and click Next. Scroll down through the services that your adapter is capable of using to the Audio Gateway option. Select it and click Next, as shown in Figure 1-45.

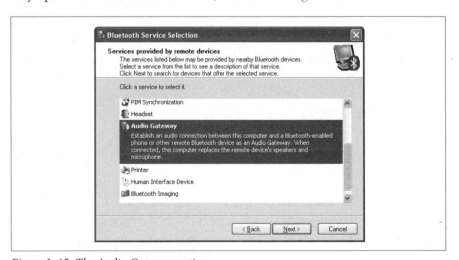

Figure 1-45. The Audio Gateway option

If you've configured your device correctly, you should see your BlackBerry along with any other Bluetooth-enabled devices within range of your adapter (see Figure 1-46)

Figure 1-46. Bluetooth-enabled devices within range

Select your device and click Next. Now check your BlackBerry—it should have detected that your computer initiated a connection with your device. There will be a dialog on your device asking for a passkey. Enter a numeric passkey and don't forget it—you will need to enter it again on your computer. After you enter the passkey on your device, Windows will prompt you that a device is attempting to *pair* with your computer. Click on the balloon icon and a dialog will appear that allows you to enter your passkey and complete the pairing of the two Bluetooth adapters. Make sure that the "Start the connection" option is selected and click Finish. Right after you click Finish, you will be prompted on your BlackBerry that your computer is initiating a connection. Choose Yes to accept the connection.

You should be able to use your computer's speakers and microphone as a headset for your device. The LED on the 7100 series devices will flash a nice blue glow when you're connected with Bluetooth, but the 7290 will give no outward indication that you're connected other than a listing of your computer in the Bluetooth section of your Options. To disconnect the session, either take your device out of the computer's range (just a few meters) or use the trackwheel to choose Disconnect from the trackwheel menu in your BlackBerry's Bluetooth section of Options.

HACK #17   Display Signal Strength as a Number

Bars are for wimps! Be different and get in touch with your inner geek by displaying your wireless signal strength in decibels.

How many bars do you have? What does the number of bars actually mean? There are entire marketing campaigns by the wireless carriers about the bars that indicate the signal strength. But how does your device determine how many bars to throw up on your display? Surely there is some type of hard measurement that the device is doing to rank your wireless connection.

It turns out there is a hard number that represents the strength of your wireless signal—and there is a BlackBerry keyboard shortcut to change the bars to display the actual number. From your Home screen, type Alt-N,M,L,L.

If you are using a 7100 series device or another Sure Type predictive typing–enabled device, you'll need to type the following sequence: Alt-B,B,M,L,L.

After typing the sequence, you should see your signal strength meter change from bars (see Figure 1-47) to a number (see Figure 1-48). You can type the sequence again to go back to the bars for your signal strength.

Figure 1-47. The signal strength meter using bars

Figure 1-48. The signal strength meter using numbers

You'll notice that this number goes higher when you coverage is poorer, and vice versa. This number is actually a reflection of what appears in the Status section of your Options program, as shown in Figure 1-49, without the dash indicating negative.

What exactly does that number mean anyway? Your signal strength is measured in units of *dBm*, which means decibels relative to 1 milliwatt. The GSM/GPRS radio inside your BlackBerry has a range of between −40 dBm at

Figure 1-49. The signal strength from your Status screen

the highest and −120 dBm at the lowest (when you display the signal strength as a number, the BlackBerry doesn't show the minus sign). Your device will display a certain number of bars given the current dBm value. Table 1-7 shows the ranges of values that correspond to the number of bars that are displayed.

Table 1-7. Signal strength versus bars

dBm range	Number of bars displayed
−40 to −77	5
−78 to −86	4
−87 to −92	3
−93 to −101	2
−102 to −120	1

Your experience may vary, but you'll probably start to notice quality problems around −100 dBm.

HACK #18 Dial Like a Pro

After becoming accustomed to cellular phone use, trying to use your BlackBerry as a phone may frustrate you at first. Here are some phone tips you'll love!

When you make the upgrade from a cell phone to a BlackBerry, you may find it difficult to get used to making calls without some of the features your phone had. If you frequently make calls on your BlackBerry, take note of some settings that can ease the process of making those calls.

Call from the Home Screen

To place a call, open the Phone application. The Phone screen appears. In the empty field at the top, you can type a phone number and then click to enter to make a call. BlackBerry 7100 device users have the luxury of dialing right from the Home screen without first going into the Phone program.

Even if you aren't lucky enough to have a 7100 series device, you can enable the same convenience of dialing from the Home screen. It saves a couple of clicks if you use the phone often!

Here's how to do it:

1. Open the Phone program. The Phone screen appears.

2. Click the trackwheel and select Options

3. Select General Options. The General Options screen appears as shown in Figure 1-50.

4. Select Dial From Home Screen. Click the trackwheel and set the option to Yes.

 When Dial From Home Screen is set to Yes, you cannot use shortcuts [Hack #3] for the applications on the Home screen.

5. Click the trackwheel and select Save.

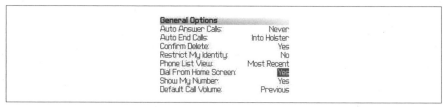

Figure 1-50. Dial From Home Screen Option

Dial Letters in Phone Numbers—(800) LET-TERS

Have you started to dial a number that consists of letters on your Black-Berry? It won't be long before you realize that you have no idea what numbers those letters translate into on a traditional dial pad so that you can place the call on your BlackBerry. Your handheld goes into number-lock mode on the Phone screen, and letters do not appear when they are entered in this mode. Don't worry, there's still an easy way to accomplish this!

To dial letters in a phone number (see Figure 1-51), you have two options:

- To type one letter, press the Alt key. Press the letter key.

- To type multiple letters, press the Right Shift key and the Alt key. Your handheld enters character mode and you can type the entire sequence of letters without having to hold down Alt. Press the Right Shift key again to turn off character mode.

 If you're using a 7100, you'll have to do a quick multitap to access the second letter on a key (an S, for example).

Figure 1-51. Dialing letters in phone numbers

Speed Dial Assignments

You can call a contact that appears on your Phone screen or in your address list (see Figure 1-52). Phone numbers also appear as links in messages and web pages, and therefore can be called through those as well.

Figure 1-52. Calling from the Phone screen

For frequently used contacts, you can assign a speed dial letter so that you can make a call by simply pressing one single key. You can also add, change, or remove a speed dial key that you have assigned.

To assign a speed dial letter:

1. Open the Phone. The Phone screen appears.
2. If the number is already listed on the Phone screen, select the phone number.

   If the number is not listed on the Phone screen, type the phone number.
3. Press and hold any unassigned key. It will be assigned to the phone number.
4. A dialog box appears with the phone number and the key you selected.
5. Click OK to confirm, as shown in Figure 1-53. The number is added to the speed dial list.

> It helps to pick a key that you can remember easily, so you may want to use the first letter of the contact's first or last name, or of their relation to you.

Figure 1-53. Confirmation screen when adding a speed dial key

View Speed Dial Assignments

To view all of your speed dial assignments, open the Phone and click the trackwheel. A menu appears; select View Speed Dial List from this menu, and the speed dial numbers will appear (see Figure 1-54).

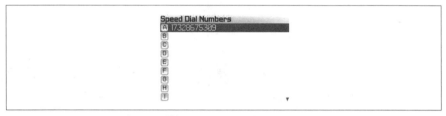

Figure 1-54. Reviewing the speed dial settings you have configured

If the contact name and number also appear in your Address Book, your speed dial list will also contain that information, as shown in Figure 1-55.

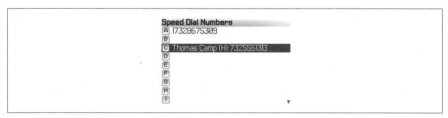

Figure 1-55. Speed dial assignment for a contact in the Address Book

—Shari Kornberg

HACK #19 Put Notes in the Call Log

Save notes from your phone calls and recall that information when you need it.

While you are on an active call with your BlackBerry, press in the trackwheel and choose "Notes" from the menu. You will be greeted with a nice clean slate of text input, ready to accept your notes, as shown in Figure 1-56.

Figure 1-56. Adding call notes to the current call

This will, of course, be most convenient when you're using a headset or speakerphone. The way I use this feature is that I jot down notes relevant to that call such as confirmation numbers or meeting notes, and save them. I can return to the Messages application, find the call in the list of calls I've placed, and find my notes there. From here, you can also copy text and put it into its final destination, whether an email, task, or appointment.

If you don't remember to take notes during your call, don't fret: you can add notes to a previous call by opening the call in the Messages list and selecting "Add Notes" from the menu. If you have decided to not show calls in your Messages list, you can magically make them appear by pressing Alt-P **[Hack #24]**. You can then surf your messages list and get these useful little pieces of metadata and put them to use. If you discussed a budget, an invoice, or something else that you may need to reference later, make a note of it in the call and you can find this information later. If you need to remember what you talked about last time you spoke to Client X, the notes will be right there in the call detail!

You can also highlight a phone call that has notes, click the trackwheel, and select Forward (see Figure 1-57) to forward it (notes and all) to an email address.

Figure 1-57. Forwarding your call notes

This lets you quickly shove the notes from that important conference call to your helpless subordinates.

—*R. Emory Lundberg*

Upgrade Your Handheld Software

#20   Find the latest version of your device's system software and install it.

Changes and new features are added to the system software by RIM at a pretty decent clip. RIM is constantly looking to fix bugs in the handheld code no matter how infrequently they crop up. Once RIM produces a new version of the handheld code, it does take a while to filter down to new handhelds, so you may find that your handheld software on your device is already a revision or two behind even shortly after you buy it.

The good news is that you can find and install the latest handheld code free of charge. Because the wireless carriers' infrastructures vary, there are versions of the handheld software per carrier. So if you use a device from Cingular, you need to monitor the Cingular download area of BlackBerry's web site for code for your device.

Find the Firmware

To ensure that you don't install software from the wrong carrier, RIM doesn't provide links to the various handheld software download areas from its site. You have to visit your carrier's site to find the link to the downloads. The wireless carrier's web sites are definitely a moving target, so to provide links here would be futile. Most of the carriers provide fairly prominent placement for the links to download the latest firmware. This can usually be found in the BlackBerry section. If you have any trouble finding the page, dial 611 (or your carrier's tech support number) from your BlackBerry and ask.

Is Your Device Already up to Snuff?

How do you know you need an upgrade? On the handheld code download page, it will give two version numbers that you can compare to your device to see if you are behind. Look for the Software Platform version number and the Applications version number on the download page for your handheld (see Figure 1-58).

On your device, go to the Options program to find the versions of the code on your handheld. To view the software platform, click on About to view the version number, as shown in Figure 1-59. The platform version on the About screen is the version number you need to compare with what's available on the web site. If the version on your handheld is less than the version on the web site, your handheld is a candidate for an upgrade.

To view software for a BlackBerry product, please select a product from the drop down menu and click Select :

BlackBerry 7290 Wireless Handheld(TM)

Submit

Software For BlackBerry 7290 Wireless Handheld(TM)

BlackBerry Handheld Software v4.0.0.274

Note: Please download the BlackBerry Desktop Software v4.0 and install it first, then install the BlackBerry Handheld software.

> Package Version: 4.0.0.274
> Consisting of:
> › Software Platform: 1.8.0.129
> › Note: The Software Platform version number can be found under Options-About screen on the handheld.
> › Applications: 4.0.0.219
> › Note: The Applications version number can be found under Options-Applications-Modules on the handheld.
> Filename: 7290E_PBr4.0.0_rel274_PL1.8.0.129_A4.0.0.219-Cingular.exe
> Size: 20.1 MB
> Posted: March 7, 2005
> Download Software
> View Release Notes - PDF
> View Upgrade Instructions - PDF

Figure 1-58. Downloading for 7290 with Cingular service

 Since I'm using the BlackBerry Simulator, the version shown in Figure 1-59 is more recent than what's currently available on the Cingular downloads area. Normally this version would be less than or equal to what you see on the web site.

Figure 1-59. Options → About

Once you have decided to upgrade your device, download the latest firmware for your model number. It will be a large download—probably around 20 MB. The handheld code is an executable that installs as a Windows application. Run through the setup just as you would for any installation on Windows.

After you install the newer handheld code on your machine, bring up Desktop Manager. The new handheld code will be detected by Desktop Manager and you'll be immediately prompted that newer system software exists, as shown in Figure 1-60.

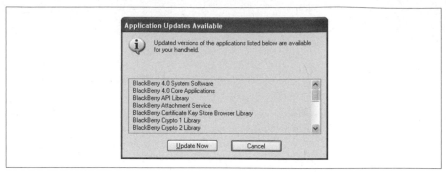

Figure 1-60. Desktop Manager recognizes the new system software

Click the Update Now button to bring up the Application Loader Wizard to install the new software on your BlackBerry. Figure 1-61 shows the new versions of the handheld code in Application Loader.

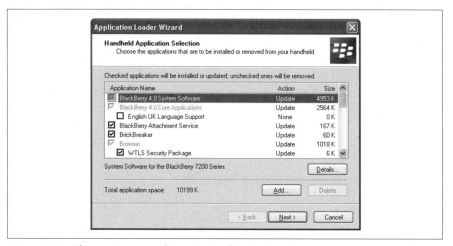

Figure 1-61. The new system software in Application Loader

You will be prompted to back up and restore all your application data associated with your BlackBerry before upgrading the system software. Although this step is optional, it is strongly encouraged because of all the customizations to your device (browser bookmarks, AutoText entries, etc.) that will be removed if you don't choose to do the backup.

 Even though your customizations will be lost if you choose not to back up and restore, any applications you've installed over the air will be retained on your device.

Type on a Bluetooth Wireless Keyboard

For those long emails, use this fold up Bluetooth keyboard to crank 'em out quicker than the best thumb typist.

While there are all types of tricks included in the BlackBerry operating system to make you type quickly with your thumbs, nothing can compare to the speed of a full size keyboard. For Bluetooth enabled BlackBerry devices, you can use a full size keyboard that makes composing those long, well thought out emails a breeze. Not only does the keyboard allow for quick typing on your BlackBerry, but you can also set up hotkey combinations that spawn certain applications.

The keyboard is made by a company called Freedom Input and costs $99. It is available for purchase at the following URL: *http://www.eaccess-estore.com/ store/catalog.asp?item=199*. The keyboard is compact and sleek. It's not a whole lot larger than your device when it's folded up. When it is unfolded, your device sits on a small stand that slides out of the top right side of the keyboard as shown in Figure 1-62.

Install the Keyboard

You'll need to do a couple of things to get your keyboard working. First, you'll need to enable the Bluetooth adapter on your device. Second, the keyboard requires that some software be installed on your device. Along with the keyboard comes a CD with a unique key printed on the back of the sleeve. You'll need to enter this key on the Freedom Input web site (*http://www. blackberrykeyboard.com/index.htm*) to register your keyboard and download the software for your BlackBerry. You'll need to use Desktop Manager and Application Loader to install the keyboard software on your device.

Pair with the Keyboard

Once the software is installed, you'll also need to *pair* with your keyboard using the Bluetooth section of your Options program. This is similar to the action taken when using your computer as a Bluetooth headset [Hack #16]. Put 2 AAA batteries in the keyboard and turn it on using the switch on the bottom-left side near the Ctrl key. In your Bluetooth settings on your BlackBerry, select "Add Device" from the trackwheel menu. As your Bluetooth adapter searches for devices within range, your keyboard should be detected and identified by the string "KEYBOARD." Select your keyboard from the list and use "0000" for the passkey. At this point, you should be able to use the keyboard to type on your BlackBerry.

Figure 1-62. The Freedom Input Bluetooth keyboard

Use the Keyboard

Your keyboard should work just as you'd expect—until you try to access the BlackBerry-specific keys such as the trackwheel. Table 1-8 shows the keys on your keyboard with which you can access the device-specific keys.

Table 1-8. Accessing the BlackBerry-specific keys with your keyboard

BlackBerry key	Bluetooth keyboard equivalent
Click trackwheel	Right arrow
Scroll trackwheel	Up and down arrows
Escape key	Left arrow
Alt key	alt gr key (beside the Fn key on the bottom row)

You'll find you won't need the Symbol key, since you have every key you'd need using the Shift key just like your computer's keyboard. There are also special symbols in yellow on some of the keys that are not found on a standard QWERTY keyboard that can be accessed using the alt gr key.

Map Function Keys

One of the nifty features of the Bluetooth keyboard is the ability to assign hotkeys to any application on your device. You can add up to 10 hotkey combinations by accessing the Bluetooth Keyboard program. Scroll down to the Function Keys section. All the available function keys can be configured by choosing a program from the list to the right of the function key. Use the Alt key to quickly choose an application. Your setting goes into effect immediately. You can also configure the repeat rate and repeat delay for the keyboard.

Email

Hacks 22–31

Rather than building an organizer and then retrofitting email support onto it, Research In Motion designed the BlackBerry for email from the beginning and only then added support for other features. Highly secure, push-based email is what has made the BlackBerry so popular in the business world. Every effort was made to allow users to efficiently process the mountain of email messages they receive daily. Email is great, but it can quickly become a burden—just go on vacation for a week without your BlackBerry and see what awaits you upon your return! There are a number of hidden features of the BlackBerry you can use to your advantage. You can clear a bunch of messages at once [Hack #22], filter your messages [Hack #24], and make your email doubly secure [Hack #28].

The hacks in this chapter will let you get through your mail quickly when you're on the go, so you won't have a pile of unread mail when you get back to the office—you'll have already read it on your BlackBerry.

HACK
#22
Clear Your Inbox Quickly

Process hundreds of messages in one fell swoop.

If you forward all your messages to your BlackBerry, there has probably been a time where you'd like to process many of the messages in your inbox at once instead of clicking through each. Sometimes an errant automated task can send you hundreds of messages that you don't need to read. At other times, you may have already read several messages using your email program on your PC, so you'd like to mark every previous message on your BlackBerry as read.

Luckily, BlackBerry offers a feature to do just this.

Process Many Messages at Once

In your Messages application, all your messages are sorted by the date they were received by your device. For each day, there will be a date heading that serves as a divider that shows you which messages were received on which day.

Use the trackwheel to select a particular date and click the trackwheel to bring up a menu. By default, the Compose Mail option will be selected. Above that selection are two options: Mark Prior Opened and Delete Prior (see Figure 2-1). These options do exactly as you'd expect—they mark all previous messages as deleted or read.

Figure 2-1. The Bulk Message functions on the menu

If you have a lot of messages on your device to mark opened, it could take some time to go through and mark each message read even if there are only a few unread messages. Your device still has to check each message on your device to see if it is read or not. To solve this, you can select multiple adjacent messages by holding down the Shift key and using the trackwheel to highlight the unread messages. This will perform the operation only on the messages you selected rather than the entire list, so it will be much faster.

Hack the Hack

You can use this hack to process the results of a search as well. Let's say an automated job that sends you email just sent you 100 messages before you were able to stop it. Perhaps they are interspersed among other important messages that you need to read. If there is some commonality in the automated messages, such as the same subject or sender, you can use the search function to confine your results to only these messages.

Select one of the errant messages in your message list, click the trackwheel, and scroll down the menu to the search section. Choose Search, Search Sender, or Search Subject to get only those messages that you want to delete. Once the search completes, you'll have just those messages that meet your criteria in the list. These search results are still divided by the date received by your BlackBerry, so you can use the steps from the previous section to confine your bulk deletion to only these messages.

If you're using a BlackBerry Enterprise Server and you've enabled wireless email reconciliation, although the Mark Prior Opened function is reconciled with your server mailbox, the Delete Prior function is never reconciled. This is to prevent an inadvertent mass deletion of emails off your mail server. To have the deletions on the device reconcile with your server mailbox, you'll need to delete the messages one by one.

HACK #23 Create Persistent Custom Searches

Use the Search function to create custom views of your mail messages. You can even assign a hotkey to execute your saved search without defining the search parameters each time.

With the mountains of email we receive each day, those messages pile up quickly. When you need to find a specific message that was sent to you a couple days ago, sifting through the message list can be a chore, especially when you've got such a small screen to work with.

If you are using the BlackBerry Web Client (BWC), you might lament how your messages *don't* pile up quickly! Instead, as soon as you get close to your miserly quota, you get a few messages about how your inbox is about to fill up, and then total silence. You may have luck calling your provider and asking for a larger quota on the BWC. Some users have had success with this, moving quotas to 20 MB or even 40 MB. You should, of course, check up on things now and then make certain you don't get an unexpected switch in the opposite direction if they make a policy change that affects all users.

Fortunately, the BlackBerry provides a nice Search function that can easily filter through your message list. There are several parameters you can use to refine your search to certain fields and message types. Perhaps the most powerful and underused feature of the Search function is the ability to save searches and make them accessible directly from the message list by typing a hotkey.

Execute a Simple Search

Let's say your boss sent you an email about the Acme Widget account, which is your biggest customer. You are not sure exactly when she sent it, but you need to look at it now, and it's not feasible to try to find it manually since your boss sends you so many emails.

From anywhere in your message list, click on the trackwheel to bring up the menu. Select Search from the menu and the screen in Figure 2-2 appears.

Figure 2-2. The Search interface

You can search your messages using a number of distinct criteria including recipient, subject, message body, folder, incoming or outgoing, and message type. Since the message from your boss included the word "widgets," you can enter that word in the Message field, click the trackwheel, and choose Search. Any messages that include the word you entered will appear in your filtered message list.

> Unless you have very few messages on your device, it will take some time to search through all of them. Be sure and look for the label in the status bar to change from "Searching..." to "Search Results" to know when the search is complete.

Search Shortcuts

When the search criteria returns still too many messages to weed through, you can use the Search interface again to specify additional parameters to further narrow your search results. One excellent feature of the Search interface is the ability to recall your previously executed search.

In the Search interface, click the trackwheel to bring up the menu and select Last, as Figure 2-3 shows. This populates the search fields with the values you used in your previous search, saving you costly keystrokes. From there you can further refine your criteria.

Figure 2-3. The Search menu

You'll also notice an option called Recall on the menu. This feature provides shortcuts to commonly used search criteria including incoming, outgoing, phone log entries, SMS messages, and voice mail messages. Choose one of these options and it will fill in the appropriate values for that search type.

Save a Search for Quick Access

One of the best features of the search interface is the ability to save a commonly used search. You assign it a hotkey that executes the search right from your message list, bypassing the search interface entirely. See [Hack #24] for some predefined search filters (such as showing all outgoing messages).

Figure 2-4 shows the Save Search screen. I've chosen to name this search "Widgets" and assign "w" as the hotkey. Use the trackwheel to choose Save from the menu. You can then press Alt-W from the message list to execute this search immediately.

Figure 2-4. Saving a commonly used search

Hack the Hack

If you have configured the BlackBerry Web Client to deliver messages from multiple accounts to your device, you can set up saved searches to access only the messages in a particular account. Set up a saved search for each account by using the account's email address and searching for any messages sent to that address. Assign a hotkey for each search, and you've got immediate access to the messages sent to each account right from your message list.

HACK
#24
Filter Messages by Type

Restrict the view to only certain types of messages in the Mail application by using these shortcuts.

As more and more messages arrive in your BlackBerry, it becomes more of a task to sort through that list to find specific items. Your BlackBerry continues to pile on by storing sent messages, sent and received SMS messages, incoming and outgoing phone logs, and voicemail notifications right alongside your email messages. Because the storage requirements for each entry in your Mail application are so small, you'll soon amass upward of 1,000 messages.

Thankfully, RIM provides some nice shortcuts to instantly filter your messages by type, allowing you to isolate a certain item you're looking for quickly. Using this hack along with [Hack #23] will let you slice and dice through your messages like the Ginsu knives you used to see advertised on late-night TV.

Use Shortcuts for Each Type of Message

The following table shows the available shortcuts to use in the Messages application. To access each shortcut, go to the Messages icon and press enter or click the trackwheel to view your message list. These shortcuts are listed in Table 2-1.

Table 2-1. Shortcuts to filter by type

Keyboard shortcut	Use to show only	Device availability
Alt-O	Outgoing messages	All
Alt-I	Incoming messages	All
Alt-P	Phone logs	All
Alt-V	Voicemail messages	All
Alt-S	SMS messages	All
Alt-M	MMS messages	All
Alt-D	Direct Connect calls	Nextel walkie-talkie devices

Make Phone Logs Disappear from Your Inbox

By default, every phone call you make and receive is logged to your Messages application. This tends to clutter your inbox. Shouldn't your inbox be just for email messages? You can tell your BlackBerry not to store these phone logs alongside your email messages.

In your Phone application, click on the trackwheel to bring up the menu. Choose the Options item from the menu. Go to the Call Logging options as shown in Figure 2-5.

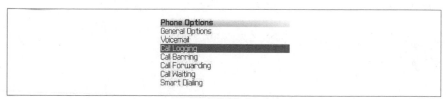

Figure 2-5. All the phone options

Here are the options for phone-log types that can appear in your messages list:

Missed Calls
   Logs only the calls you weren't able to pick up before they went to voicemail.

All Calls
   Logs everything, including incoming, outgoing, and missed calls.

None
   Doesn't log any type of calls to your message list.

Before you go choosing to list all types of phone logs alongside your messages, consider the following. When you choose to log none of the phone call types, they actually still exist in the message list—they are just hidden from your view. You can still use the shortcuts from this hack to view them from the Messages application. All the convenience of being able to view the phone logs without them cluttering up your mailbox—what a bargain!

HACK #25 Use an Auto Signature

Customize your signature and have it automatically appear at the end of each message you send.

When you first started using your BlackBerry, you probably noticed that with each message you sent from it, some text was appended to the end of your message. When you originally set up your BlackBerry, the default text is something similar to "Sent from my BlackBerry Wireless Handheld."

While this is certainly convenient as a Research In Motion marketing tool, it would be nice to be able to change it to whatever you like. It turns out you can.

Non-BWC Users

If you are using a BlackBerry Enterprise Server or the Desktop Redirector for your mail delivery, you can adjust your auto signature in Desktop Manager. Double-click on Redirector Settings in Desktop Manager to bring up the properties. Under the Auto signature section of the General tab, customize your signature to whatever you like (see Figure 2-6).

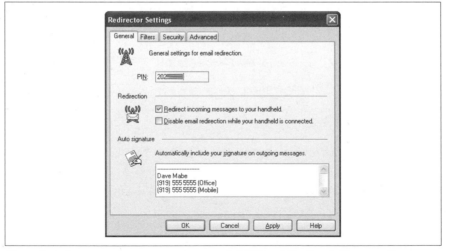

Figure 2-6. Customizing your auto signature for BES and Desktop Redirector clients

BWC Users

If you use the BlackBerry Web Client for email redirection to your device, you need to go through the link for your provider's BWC web interface. Once you've logged into the site, click on the Options link at the top of the page. This brings you to a page, shown in Figure 2-7, which looks very similar to the property page for BES users. BlackBerry Web Client users also get to use the "Include auto signature in outgoing messages" option to toggle this feature on and off.

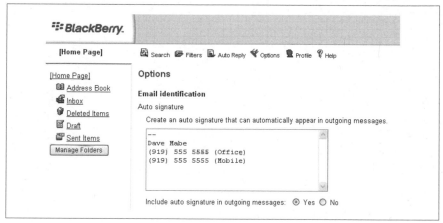

Figure 2-7. Customizing your auto signature in the BlackBerry Web Client

Hack the Hack

If you've used the multiple mail hack [Hack #31] and you're using the BlackBerry Web Client service in addition to a BlackBerry Enterprise Server, you can use a different auto signature for each service you use. Just configure your auto signature in both places, and mail you send through each service will be appended with its appropriate signature. This is helpful if you use an official, company-mandated auto signature for company mail sent through a BES, but you want to let it all hang out with your fancy ASCII-art personal signature when you send emails from your personal account.

You can also use AutoText [Hack #7] to insert custom signatures on the fly.

HACK
#26

Send a Message to a Group of Users

Do you have a group of people you'd like to communicate with regularly? Here are some tips for sending email or peer-to-peer (PIN) messages to several addresses.

If you use *distribution lists* (DLs) to address email in the enterprise, you can use those same DLs to address email to groups of people from your BlackBerry.

Also, it is pretty simple to create and send a PIN message to a person [Hack #27]. You can also include the PINs of several people, if you'd like the same message to go to multiple recipients. That works for an occasional message, but if you have a particular group of people to which you would like to send PIN messages frequently, you'll find yourself getting tired of adding the individual recipients each time you create a message. You cannot use an Address Book group to send PIN messages, but you can still accomplish the task in other ways!

Use Distribution Lists

Any distribution list can be added to your Outlook Contacts in the same manner you would add any other entry from your enterprise *Global Address List* (GAL). Simply select the name in the GAL or email message, right-click, and select Add to Contacts. When your contacts are synchronized with the Address Book on the handheld, those entries will be copied to your handheld, and therefore can be used to address email from the handheld. Alternatively, if you have a message that was addressed to a distribution list, that has been received on the BlackBerry handheld, you can also use that to save it in your Address Book for future use. Here's how:

1. Open the messages list.
2. Open a message that has been addressed to one of the distribution lists you would like to add to your Address Book.
3. Scroll up to the To line that contains the distribution list.
4. Click the trackwheel and select Add to Address Book.
5. The Address Book will open and display the new entry with the email address of the DL prepopulated.
6. Complete the remaining fields and select a name for the entry so that you can easily identify the DL in your Address Book.
7. Click the trackwheel and select Save.

Now you can use this DL in the To line of any message you address!

Create a Boilerplate PIN Message

Next, you need to create a sample PIN message and address it to a group of users. But don't send it just yet!

1. Open the Messages application on the handheld.
2. Click the trackwheel and select Compose PIN. The Address Book application screen appears. Select the first user you would like to include in this PIN message, click the trackwheel, and select PIN *username*. The Message screen appears. Enter a subject line such as "TEST PIN" for now.

3. Click the trackwheel and select Add To for each additional recipient of this message. A sample message addressed to several PINs is displayed in Figure 2-8.

4. Now that your message is properly addressed, click the trackwheel and select Send. The Messages screen appears, and you can see your message with a checkmark to the left indicating that it has been sent.

5. Select the sample message, click the trackwheel, and select Save. This will move your sample message to the Saved Messages application.

 If your handheld decides that it needs to delete messages to free memory, it will not delete saved messages.

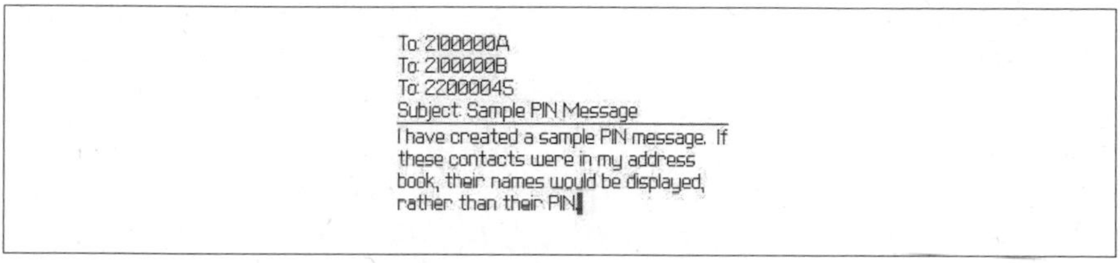

To: 2100000A
To: 2100000B
To: 22000045
Subject: Sample PIN Message
I have created a sample PIN message. If these contacts were in my address book, their names would be displayed, rather than their PIN

Figure 2-8. Sample PIN message addressed to three recipients

Use Your Saved Message to Send a PIN Message

Now you have a saved message in your saved messages application that you can use over and over to send a message to the users who were added as recipients. Simply open the Saved Messages application and locate your sample message.

Then follow these steps:

1. Click the trackwheel and select Open. The sample message opens.

2. Click the trackwheel to select Reply to All.

3. Click the trackwheel once again to select Delete Original Text.

4. Once the message is open, modify the Subject line so that it is appropriate, and enter the body of the message you are composing.

5. Once you are satisfied with the subject and body of your message, click the trackwheel and select Send.

You can use this sample message over and over again. You can also modify this message to add or remove recipients at any time, or to change the PIN of a recipient.

—*Shari Kornberg*

HACK
#27 # Send a Message Directly to Another BlackBerry User

Do you regularly communicate with a particular person or several people who also have BlackBerry devices?

BlackBerry handhelds can send and receive email messages that are routed through your existing email account. You can also use your handheld to send and receive peer-to-peer, or *PIN*, messages. All BlackBerry devices are assigned a unique personal identification number (PIN), which is used to identify the handheld on the network. Provided you know the PIN of another person's handheld, you can add it to your Address Book and use it to send a PIN message directly to that person.

To locate your handheld PIN, open the Options application on the device. The Options screen appears; select Status. The PIN field within Status displays your handheld's PIN, as shown in Figure 2-9.

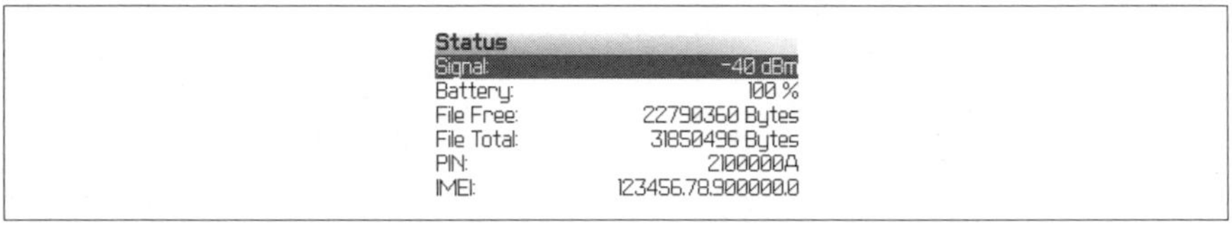

Status	
Signal:	-40 dBm
Battery:	100 %
File Free:	22790360 Bytes
File Total:	31850496 Bytes
PIN:	2100000A
IMEI:	123456.78.900000.0

Figure 2-9. Use Options → Status to locate your handheld PIN

PIN-to-PIN messages are not routed through an email account—they bypass your mail server and do not appear on the desktop. They are routed directly from one handheld to another through the wireless network using the devices' unique PIN numbers. An advantage of this is that it provides a measure of communications fault tolerance, an alternate way to route a message in the event the mail system is not available.

PIN messages can also be helpful during troubleshooting. Send a PIN message to your own device or another, and you can verify that it can be contacted on the network and the device is working!

It is simple to create and send a PIN message to a person. You can also include the PINs of several people if you'd like the same message to go to multiple recipients.

Set Up the PIN Fields of Contacts in Your Address Book

You will need to first obtain the PINs of each person with whom you would like to exchange PIN messages. It's best to include the PIN in your Address Book entry for the contact using the PIN field. This way you will be able to easily identify who you are communicating with when you receive a PIN message. Otherwise, you may receive a PIN message and won't be able to tell who the sender was! It is important for folks to provide you with their PINs and notify you if they ever change or upgrade to another device.

> Since PIN numbers are unique to the handheld, you will always want to be sure that you have the current PIN of the person with whom you are communicating. If not, your message may arrive on the handheld of a person for whom the message was not intended! So if your friend casually mentions that he's selling his current BlackBerry on eBay, make a note [Hack #47] to get his new PIN when he gets his new BlackBerry.

Open the Address Book application on the handheld. Select and edit the contact for whom you'll be adding a PIN. Scroll down to the PIN field and enter the eight-digit PIN. A PIN may contain both numbers and alpha characters. Click to Save and Close the Address Book.

Set Up the PIN Fields of Contacts Using Outlook

If you are not too keen on entering a lot of information on the handheld, and you have several PINs you'd like to include in your Address Book, you can also enter them using Outlook. When your Outlook Contacts are synchronized, your handheld will be updated as well.

1. To enter PINs using Outlook, open your Outlook Contacts, and then select the entry you would like to modify.
2. Double-click the entry to edit it, and then select the All Fields tab.
3. For the Select From field, use the drop-down arrow to select Miscellaneous Fields.
4. The field named User Field 1 maps to the PIN field for that Contact on your handheld.
5. Enter the handheld PIN in the Value field, as shown in Figure 2-10.
6. Repeat for each PIN you would like to enter.
7. Synchronize your Contacts, and the PINs will be updated on your handheld.

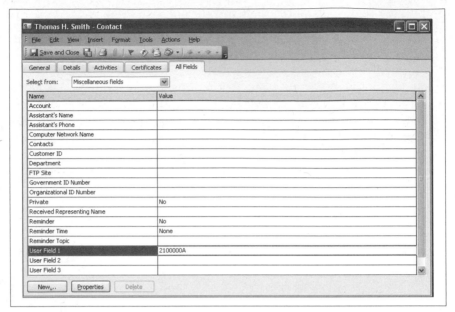

Figure 2-10. Outlook Contact entry with PIN field populated in User Field 1

Create and Send a PIN message

To create and send a PIN message, do the following:

1. Open the Messages application on the handheld.

2. Click the trackwheel and select Compose PIN. The Address Book application screen appears. Select the user to whom you would like to send this PIN message, click the trackwheel, and select PIN *username*. The Message screen appears. Enter the Subject line and message.

> You can also select Use Once from the Address Book and type a PIN number.

3. Click the trackwheel and select Add To if you would like to add additional recipients to this message.

4. Now that your message is properly addressed, click the trackwheel and click Send. The Messages screen appears, and you can see your message with a checkmark to the left indicating that it has been sent.

When your PIN message is delivered to the recipient, a D appears with a checkmark in the messages list.

You can also use short message service (SMS), to communicate with other users that have SMS-compatible phones. An SMS-compatible number is a phone number that your service provider enables for SMS. Typically there will be a per-message fee for SMS messages, so PIN messages may save you some money since they may be included in an unlimited data plan. Check with your provider. Also, an SMS message can take quite a bit of time to deliver (usually minutes, but sometimes hours), and if it can't be delivered within 24 hours, it's silently dropped like so many bits into a bucket.

All handhelds share a common encryption key that is loaded during manufacturing. PIN messages are encrypted with Triple-DES; however, the key to decrypt the message is available to everyone with a BlackBerry handheld. Therefore, PIN messages should be considered scrambled, but not encrypted. For security reasons, if you are sending proprietary or sensitive corporate information, you will want to use the mail account that routes through the BlackBerry Enterprise Server.

—*Shari Kornberg*

HACK #28 Send and Receive S/MIME Encrypted Emails

By default, you won't be able to view any S/MIME messages on your BlackBerry. Here's how to set up your device to be able to view them.

While all communication between your device and your server is encrypted, the emails that you compose and receive end up as plain text when they're ultimately sent across the public parts of the Internet. S/MIME solves this problem by providing a way to send signed and encrypted emails across the Internet. S/MIME ensures that, in emails that are digitally signed, the sender is who she says she is, and the content wasn't altered in transit. Encrypted emails are encrypted using the recipient's public key, guaranteeing that the content can be unencrypted only by the recipient's private key.

By default, the BlackBerry can't read emails sent using S/MIME. However, if you're using a BlackBerry Enterprise Server, there is an additional software package you can purchase from RIM that allows you to send and receive S/MIME encrypted emails. You can even wirelessly retrieve the digital certificates of users to whom you'd like to send encrypted email via an LDAP server.

Install the BlackBerry S/MIME Support Pack

When you originally installed your BlackBerry Desktop Manager, there was an optional component called Certificate Synchronization that you may or may not have installed. If you did not, you'll need to go through Add/ Remove Programs in Control Panel and modify the BlackBerry Desktop Manager installation to ensure that the option is selected (see Figure 2-11).

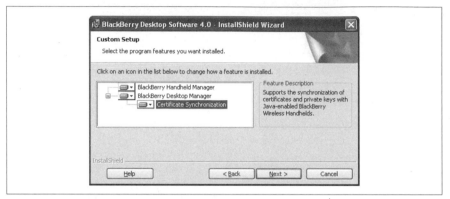

Figure 2-11. The Certificate Synchronization custom option

After you have installed the Certificate Synchronization option, you'll have another icon in Desktop Manager (see Figure 2-12), where you'll configure various options related to S/MIME. You'll also need to go through Application Loader to load the updated S/MIME libraries on your device. This will add a couple items into your device Options and set up your mail program with the ability to send messages using S/MIME.

Figure 2-12. The Certificate Sync icon in BlackBerry Desktop Manager

You'll need to purchase another software package directly from your RIM representative called the BlackBerry S/MIME Support Package. This is a separate installation that you need to run on your computer. It will install the libraries on your BlackBerry that allow you to send and receive S/MIME email messages.

BES Configuration

The BlackBerry Enterprise Server that your device is homed on will have to have S/MIME enabled as well. In the BlackBerry MMC, ensure that the "Support S/MIME encrypted messages on this server" option is checked, as shown in Figure 2-13.

Figure 2-13. Enabling S/MIME on your BlackBerry Enterprise Server

Import Your Personal Certificates to Your Device

Once you get the software loaded and make sure your BlackBerry environment is set up for S/MIME, you will need to load your certificates in to your device's key store. Bring up the certificate synchronization tool by double-clicking on the Certificate Sync icon in Desktop Manager.

First off, you'll need to add your private key to the BlackBerry key store. Click the Import Certificate button (see Figure 2-14), browse for your private key certificate, and select it. It will be imported into the key store and then synchronized with your device.

Import Other People's Certificates over USB

To send encrypted email to other people, you need to encrypt the contents with their public key. This encryption takes place on your BlackBerry, so you will have to get those users' public keys onto your device.

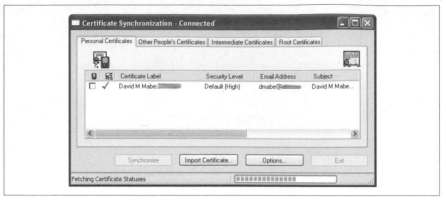

Figure 2-14. Importing a private key

There are a couple ways to do this:

- Search and import the certificates from an LDAP server that houses the certificates.
- Import the certificates by browsing to a certificate file on your local filesystem.

The first option is certainly the easiest, if it's available to you. Most Public Key Infrastructure (PKI) deployments have LDAP interfaces for accessing other users' public keys. This provides a convenient way for users to send encrypted email to a user with whom you haven't explicitly exchanged keys. Making users' certificates available via LDAP alleviates the key exchange problem that plagued PKI in its infancy.

To import a certificate that is stored on an LDAP server, you will need to define the settings describing the LDAP connection. Click Options in the Certificate Sync tool and go to the LDAP Servers tab. Click the Add button to add a new LDAP server.

Enter a descriptive name for the Friendly Name field in the dialog box, as shown in Figure 2-15. The Base Query defines the search base, which will be different for each domain that the LDAP server is able to search.

If your certificates are stored in an Active Directory domain, then, by default, you will have to set the Authentication Type to Simple and use your Windows domain credentials to bind to the LDAP server.

> While you can bind anonymously to a Windows Active Directory domain, your search results will come back empty. Be sure and specify an account to use to bind to the LDAP interface on a Windows domain controller so you'll have access to search for certificates.

Figure 2-15. Defining the LDAP connection properties

Once you've defined an LDAP connection, click OK to return to the Certificate Sync tool. Go to the Other People's Certificates tab to import other certificates to sync with your BlackBerry. You can then use the Find in LDAP button to search for certificates stored in the LDAP server you specified. When you find a certificate you would like to have available on your Black-Berry, select the entry from the search results and click the Mark for addition button, as shown in Figure 2-16.

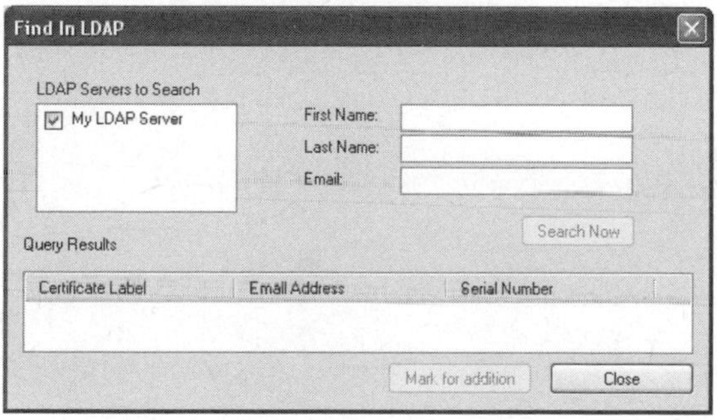

Figure 2-16. Adding another user's certificate using LDAP

Once you've added the certificates you'd like to have available on your Black-Berry, click the Synchronize button to load the certificates onto your Black-Berry's key store through the USB cable. Your key store on your device is secured by a password that is defined when you use the Certificate Sync tool (shown in Figure 2-17) the first time. When you synchronize your certificates in Desktop Manager, you will be prompted for this password. This is different from the password for unlocking your device (which you should have set!), which you are also prompted for when you synchronize. The first time you do this, it can be confusing to determine which password you're being prompted for. Read carefully to ensure you are entering the correct one.

Figure 2-17. Setting your new key store password

Import Certificates via LDAP Wirelessly

You can also do certificate queries and lookups wirelessly from your device. To enable this functionality, your BES administrator will need to set up some LDAP settings similar to what you set up in your client. The configuration screen is shown in Figure 2-18. When these settings are changed, the MDS service will need to be restarted.

Figure 2-18. MDS LDAP settings for wireless certificate lookups

Once your BlackBerry Enterprise Server is set up with the proper LDAP settings, you'll be able to retrieve the certificates for a user with whom you haven't already exchanged keys. When you are composing a message, choose the type of message you'd like to send, and if there isn't a certificate for the recipient on your device, it will automatically do a wireless lookup and try to retrieve the recipient's public key.

In the documentation that comes with the BlackBerry S/ MIME Support Pack, it incorrectly states that you can use S/ MIME with any version of BES 3.5 or greater. If you are using an Active Directory domain controller for your LDAP source, the wireless lookups won't work unless you have BES 4.0 or higher.

Once your server is configured, you can perform wireless certificate searches from your device (see Figure 2-19). You can add certificates as you compose a message on your device or you can import certificates into your device using the Certificates program in Options.

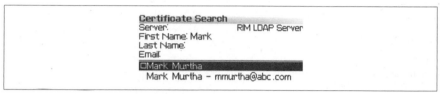

Figure 2-19. Searching for a certificate from your device (image courtesy of Research In Motion, Limited)

HACK #29 Six Ways to Check Your Gmail

There are a variety of ways to check your Gmail account from your device.

Maybe you've had a Gmail account since Gmail invites were going for $50 on eBay; maybe you're a new user. So what is the best way to access it from your BlackBerry? Well, it depends. Are you a BES user? Does your IT policy restrict access to the BlackBerry Web Client? Does your carrier supply only a severely limited WAP browser? Even if you're in a high security environment, chances are you can access the world's best free email service. Here are six ways to access your Gmail from your BlackBerry.

Redirect Gmail to Your BES Account

If you are a BlackBerry Enterprise Server user, the quickest and easiest way to get your Gmail delivered to your device is by setting up your Gmail account to forward all your mail. In your Gmail account, go to Settings → Forwarding and POP (see Figure 2-20). Select the option for Forward a copy of incoming mail to and fill in the email address of your BES account. You can choose to keep a copy of each email message in your Gmail inbox, archive a copy, or trash the Gmail copy. If you are going to access your Gmail using the other methods in this hack, go ahead and keep a copy of the message in your inbox.

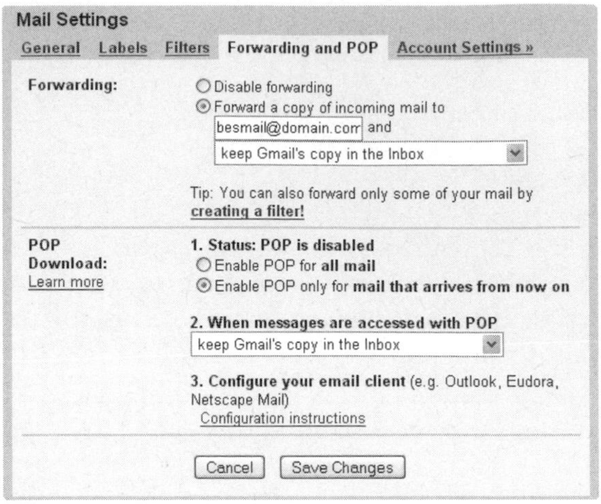

Figure 2-20. Gmail's Forwarding and POP settings

Set Up POP and Use the BWC

If you feel a little queasy sending your personal Gmail messages to your corporate email account, you can enable your Gmail account for POP access and use the BlackBerry Web Client to access it. In your Gmail, go to Settings → Forwarding and POP. Select the checkbox next to "Enable POP only for mail that arrives from now on." Then set up your BlackBerry Web Client with your Gmail address as the username and your Gmail password. There's even a walk-through on Gmail's Help Center specifically for setting up the Black-Berry Web Client to access your Gmail.

> Because the BlackBerry Web Client has to poll your mail-boxes before delivering mail to your device, there will be delays of up to 15 to 20 minutes before receiving some messages. However, when using a BlackBerry Enterprise Server with Microsoft Exchange, you'll get your messages to your device almost instantly. This is because the BES uses MAPI connections with Exchange and receives instant new mail notifications—a feature that sets Exchange apart from its competitor Lotus Domino.

Use a Browser to Access Gmail's HTML Version

It used to be that if you wanted to access Gmail with an older browser on your computer, you were out of luck. Gmail uses some complex JavaScript techniques that require a relatively modern browser on your desktop. Although supported in BlackBerry 4.0, the JavaScript implementation in the BlackBerry Browser isn't hearty enough to handle Gmail's code.

Luckily, Google has fairly recently created an HTML-only version of Gmail. This allows you to access your Gmail account with the built-in BlackBerry Browser or a third-party browser like WebViewer [Hack #66]. Just point your handheld browser of choice to *http://gmail.google.com*, and Gmail will do some JavaScript sniffing to determine that your browser is best suited by their HTML version. You'll know you're accessing the HTML version by the notice at the top of the page, as shown in Figure 2-21. WebViewer also provides nice access to the HTML version of Gmail.

Figure 2-21. Gmail's HTML version in the BlackBerry Browser

Use EmailViewer via Gmail's POP Interface

Just as you would access your Gmail using POP from your computer's email client, you can use the EmailViewer application [Hack #67]. You'll set up the connection much as you would any standard POP mail settings. Gmail's help pages include an article for setting up a generic email, available at *http://gmail.google.com/support/bin/answer.py?answer=13287&topic=194*.

Note that EmailViewer doesn't support Gmail's implementation of SMTP over SSL. Just leave the SMTP settings blank in EmailViewer to use ReqWireless's mail relay service.

Use gmail-mobile to Access Gmail via WAP

If you're restricted to using the carrier-provided WAP gateway, you can use a nifty PHP application that provides a WAP interface to Gmail. There are various sites that have this software installed already. If you choose to use one of them, you'll need to realize that your Gmail username and password are easily accessible by the site hosting the gmail-mobile application. Luckily, the gmail-mobile project is open source software, so you can download the source and install it on your own PHP-enabled Apache web server. The interface provided by gmail-mobile, shown in Figure 2-22, turns out to be the most convenient and usable of the solutions that don't involve using the BlackBerry Mail app through a BES or the BWC. It even provides support for labels and shows your overall Gmail storage usage.

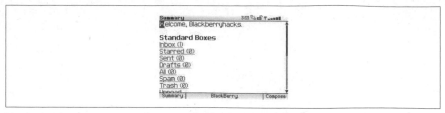

Figure 2-22. The WAP interface provided by gmail-mobile

See Also

- gmail-mobile (*http://gmail-mobile.sourceforge.net*)
- "Use Gmail as a Spam-Catcher" **[Hack #48]**

HACK #30 Use Filters to Control Message Delivery

Use BlackBerry's built-in filtering capabilities to filter messages or classify certain ones as Level 1 messages.

Imagine you have just sent your manager an urgent request and you know he'll be replying sometime soon. You would like to be notified the instant he replies to your message, given the importance of the issue. You could stare at your BlackBerry, waiting patiently for those transmission arrows to appear in the top-right portion of your screen, indicating a message is on its way. But you've got other work to do!

You can use a filter in conjunction with your device profiles to automatically notify you when his reply arrives, so you can get on with your life. Using this hack will let you avoid constantly checking your device for new messages—an action that brings down the "spouse approval factor" for the BlackBerry.

Set Up Custom Filters

If you're a BlackBerry Enterprise Server user or you use the BlackBerry Desktop Redirector, you can customize the filters that are applied to your messages in Desktop Manager. Double-click on the Redirector Settings icon and go to the Filters tab. Click on the New button to configure a new filter. This brings up the Add Filter dialog box, as shown in Figure 2-23.

BlackBerry Web Client users can configure new filters on the BWC web site by clicking the Filters link at the top of the page, as shown in Figure 2-24.

Enter a descriptive name in the Filter Name field. Use the conditions in the "When a new message has the following conditions:" section to narrow down the messages that you'd like to tie to a particular action. These actions will be "ANDed" together—in other words, the filter action will apply only when all the conditions are satisfied.

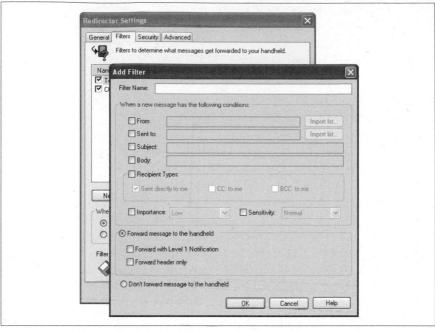

Figure 2-23. Adding a filter

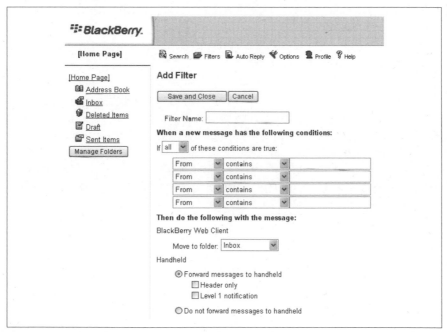

Figure 2-24. The Add Filter screen for BWC users

When using the From or Sent to fields to specify conditions, you can manually type a standard email address in the field or use the Import list button to pick the email addresses from your corporate Global Address List or your Outlook Contacts folder. To specify your manager's email for this example alert, type it in the From field.

> In the From or the Sent to fields, use a semicolon to separate multiple addresses. When multiple addresses are specified on either of these fields, they are "ORed" together—so the condition will be met if any of the addresses entered appear in its respective field in an email message.

Use the Recipient Types field to isolate messages that are sent to you in a certain way. Use the Sent directly to me option to require that your email address be specifically specified on the To line before the action is taken. This is useful for isolating messages that are sent directly to you from messages that arrive via a distribution list. In this example, you are probably more likely to want to have your device alert you when a message is sent directly to you.

Set the Action for the Filter

If the criteria you've specified in the conditions section are met, then you can choose how the messages are handled. To have your device categorize them differently for alerting, use the Forward with Level 1 Notification option. This allows you to instruct your device to alert you in a specific way for these types of messages.

If you don't have an unlimited data plan, you may want to make use of the Forward Header Only or the Don't Forward Message to the Handheld options. The Forward Header Only option sends only the header fields to your device at first. Once you open a message that's been sent this way, you will see the To, Sent, From, and Subject fields of the message, but the body of the message will not appear. Instead, you'll see the text More Available in the body along with the number of bytes in the body. Click the trackwheel once to bring up the menu and choose More or More All to retrieve the body of the message.

> Notice the difference between the Forward Header Only option and the Don't Forward option: one option forwards minimal information to your device and the other option gives you no indication at all on your device that a message was received in your inbox.

Use Your Device Profiles to Set Up Alerting

On your device, you can set up special alerting when messages that you've categorized as Level 1 are sent to your inbox. From the Home screen on your device, choose Options, and then choose Profiles. To edit a profile, use the trackwheel to bring up the menu and choose Edit, as shown in Figure 2-25.

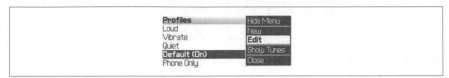

Figure 2-25. Editing your profiles

Use the trackwheel to select Level 1 messages and press Enter to edit the alerting associated with these messages. Notice there are two sections, shown in Figure 2-26: one for when the device is in the holster and another for when the device is out of the holster. There is a small magnet in the holster that the device uses to detect whether it's in the holster. Having separate alerts for in and out of holster is convenient for specifying an audible alert when your device is out of its holster and perhaps sitting on your desk or counter.

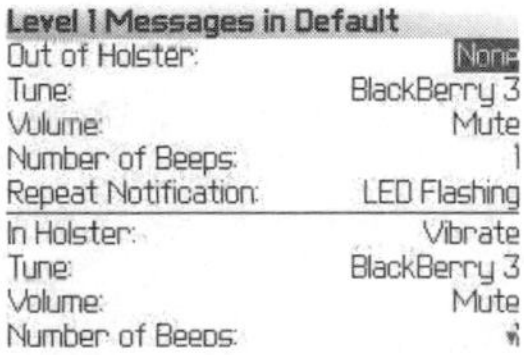

Figure 2-26. Customizing alerts for Level 1 messages

These options should be self-explanatory for the most part, except perhaps the Repeat Notification option. This option lets you specify what the device does if you've received a message and then not read it yet. Perhaps you've laid your device down to exercise or sleep (surely you don't wear your Black-Berry doing either of these!) and you return. Using the Repeat Notification option will let you be alerted when an important message hasn't been attended to.

If you're using a BlackBerry Enterprise Server or the Desktop Redirector for delivery, then by default, your messages won't be forwarded to your device when it's cradled and connected to Desktop Manager. This behavior can be changed on the General tab in your Redirector Settings. I have received several phone calls in my day from users who think the BES server is down because they aren't receiving the test messages they are sending themselves when setting up their custom filters.

HACK #31 Get Mail from Multiple Accounts

Why use your unlimited data plan for only one email account?

So you've just got your BlackBerry up and working and you've instantly fallen in love with it. You are excited about how much more productive you can be with it just by having access to your email while not sitting behind your desk. You'd really like to have the same functionality for your personal email account in addition to your corporate email, which you've configured your device for. You can always check your personal account via the BlackBerry Browser, but that interface is clunky and not nearly as usable as the built-in mail program.

You're so happy with your current device that you've even considered getting a second device so you can access your secondary email account. Well, don't go spending any money on another device just yet.

Many BlackBerry users don't realize that their device is able to send and receive messages from multiple mail accounts. There are multiple ways to have mail delivered to your BlackBerry without installing a third-party mail program [Hack #67] and without buying an additional device and service.

Use the BlackBerry Web Client

If you are a BlackBerry Enterprise Server (BES) user, you may not realize the extent of your device's usefulness. Fairly recently, RIM introduced the BlackBerry Web Client, or *BWC*, which allows you to send and receive email from accounts you configure without running any software on your computer (such as the BlackBerry Redirector). If you already use the BlackBerry Web Client for email, did you know you can access up to 10 email accounts with it?

Although the service runs on RIM's infrastructure, to access the BlackBerry Web Client, you will need to go through your wireless provider's web site. Table 2-2 shows each provider's portal to the BWC, each with the provider's brand and logo weaved into the site.

Table 2-2. Major U.S. providers and their BWC addresses

Provider	BWC URL
Cingular	*https://webclient.blackberry.net/WebMail/Window.jsp?site=cingular*
Nextel	*https://webclient.blackberry.net/WebMail/Window.jsp?site=nextel*
Sprint	*https://webclient.blackberry.net/WebMail/Window.jsp?site=sprint*
T-Mobile	*https://webclient.blackberry.net/WebMail/Window.jsp?site=tmo*
Telus Mobility	*https://webclient.blackberry.net/WebMail/Window.jsp?site=telus*
Verizon	*https://webclient.blackberry.net/WebMail/Window.jsp?site=verizon*

Once you create a BWC account, you can add your email accounts to the service. You can configure up to 10 accounts, but when you send from your account using the BWC, you'll only be able to use a single From address. You can, however, have different Reply-To addresses that you can choose from on a per message basis. Figure 2-27 shows the interface for adding email accounts to the BWC.

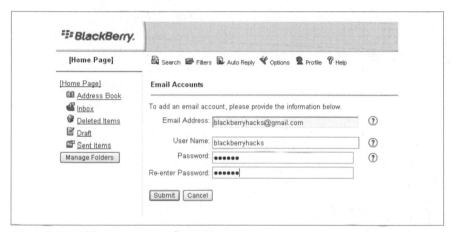

Figure 2-27. Adding accounts to the BWC

Forward Mail to Your BES Account

If your company has restricted access to the BWC using IT policies [Hack #99], you can always simply forward messages to your BlackBerry Enterprise Server account. There are a variety of ways to set up your personal account to automatically forward mail to the email address that your BES is configured to use.

.forward file
> On most Unix-based mail servers, you can set up a *.forward* file in your home directory to forward all mail to a specified address. Use your favorite text editor to edit the *.forward* file and add the address to which

you'd like to forward your mail as a single line in the file. After saving the file, your mail server will forward any incoming mail from that moment forward to the address you specify.

procmail

Some Unix-based mail servers give you access to *procmail*, a nifty utility for doing complex processing of email messages. When configured, *procmail* runs a set of rules (or *recipes* as they are commonly called) on each message that arrives. There are countless ways you can process your email with *procmail*—you can run your messages through a spam or virus checker, check for the size of a message, indicate whether it has attachments, add custom headers, etc. You can set up complex recipes that forward mail to your BES only when certain specific conditions apply (for example, when an email has been sent directly to you and not through a mailing list or a CC). For more information, see the *procmail*, *procmailrc*, and *procmailex* manpages.

sieve

Sieve is a popular server-side filtering language similar to *procmail* that is included in the Cyrus IMAP server. It allows users to create rich filters for email that live on the IMAP server. Some email clients allow users to directly manipulate these rules as well, making it very elegant. For more information on *sieve*, see RFC 3028 (*ftp://ftp.rfc-editor.org/in-notes/rfc3028.txt*).

Rules Wizard

If your account is hosted on a Microsoft Exchange Server and you use MAPI to access it, you can use the Rules Wizard within Outlook to set up server-based rules. Although not nearly as flexible as *procmail*, you can set up similar criteria to forward email based on different properties of the message.

Other ways to forward

There may be other ways to forward mail from your account. For example, Google's Gmail service lets you set up rules via the web interface to selectively forward messages to your BES account (see Figure 2-28). Of course, with Gmail there are a variety ways to check your mail with your BlackBerry **[Hack #29]**.

Figure 2-28. Setting up forwarding in Gmail

BlackBerry Desktop Redirector

The BlackBerry Desktop Redirector is software that runs on your computer that acts as your own personal BlackBerry Enterprise Server. You configure it to point to a specific MAPI profile, and it will check that profile and forward mail to your BlackBerry from that account. If you can configure an email account with Outlook, then you can have the Desktop Redirector use it.

Although at first this seems like a useful alternative, there are a few reasons this is probably the least attractive option. First, the Desktop Redirector runs on your personal computer, so your computer will have to stay on and logged in all the time for email delivery to occur. This may or may not be an option. In addition, your personal Internet connection is probably far less reliable than the BlackBerry Web Client that is fully supported and monitored and has redundancy built in. Also, a BlackBerry Enterprise Server that is run by a corporation likely has redundant network links, service monitoring, and fault alerting integrated into the service. Third, if you attempt to run the Desktop Redirector on a computer on your intranet, you may not be able to make the required TCP connection to RIM's SRP network on port 3108. On secure networks with firewalls, this connection is likely not allowed by default, and you'll have to convince the firewall administrators to allow it—a task that is usually easier said than done in most companies.

All these methods can be used simultaneously (assuming none are specifically restricted by your company), so you can pick and choose which methods are best for you.

CHAPTER THREE

Games
Hacks 32–36

The BlackBerry as a gaming platform? Although the BlackBerry is optimized for the suit-and-tie business crowd, newcomers are pleasantly surprised at the graphics capabilities of the device. Currently, Magmic Games has a big lead in the BlackBerry gaming market. There are even some games that you can play against online opponents. While Magmic pushes the limit of BlackBerry games, there are plenty of free games you can download and install over the air in seconds. You could even use the device simulator [Hack #93] to play the games.

HACK #32 Play Texas Hold 'Em

Quit your day job and start playing everyone's favorite flavor of poker for a living!

If you want to play the best games for the BlackBerry, look no further than Magmic Games (*http://www.magmic.com*). Their top-selling game is so addicting that you may have trouble going back to using your device for email. Texas Hold 'Em King 2 is the king of poker games, allowing you to play against virtual opponents offline or live players in its online mode.

The program can be installed from Magmic's download page, which is formatted for your BlackBerry Browser, at *http://bb.magmic.com*. Just choose the over the air install, and away you go. If you'd like to use Application Loader to install the game, you can download all Magmic's color or black-and-white games in one fell swoop at *http://www.magmic.com/download.php*.

Ante Up!

The program comes with a seven-day trial. After the seven days have elapsed, you can purchase a key at Magmic's web site for $9.95. To play an offline game against the AI opponents on your BlackBerry, choose New Game from the main menu, as shown in Figure 3-1.

Figure 3-1. Starting a new game

If you've never played casino style Hold 'Em, the object is to take everyone else's money—and if you've ever played in an actual casino, it doesn't take long to figure this out. Most people stay in with bad hands far too long. If you don't have the cards, go ahead and fold. (No avoiding the obvious reference to a certain Kenny Rogers song.) Just as with many gambling games and even the stock market, your goal should be to *not lose money* as much as it is to actually make money.

What's great about Texas Hold 'Em King is that most of the game is handled automatically. You don't have to hit any keys until you actually have to make a decision. This is quite nice because the game goes just slowly enough for you to recognize what the other players are doing, yet it's fast enough to keep your interest.

When it's your turn to bet, your lowest possible play to stay in the game is selected. If you want to raise the bet (or fold), use the trackwheel to increase your bet. If you want to fold, scroll down on the trackwheel and choose Fold. Click the trackwheel or use the Enter key to make your selection.

 You'll notice that if no one has increased the bet, you'll get the option to Check without an option to fold. Just like if you were at a real casino, you would be happy to stay in the game without risking any more money to see another card turned over. So even if you have a bad hand, you'd want to check, if possible, to see if your hand improves.

Considering your BlackBerry device is no Xbox, the graphics are fantastic (see Figure 3-2). The opponents line the top of the screen, and you can see the amount of (virtual) money each has left. The yellow token indicates the dealer. In the first game, everyone starts out with $500, and there are $5 and $10 blinds. As you continue to win games (not just individual hands), you graduate to games with higher stakes.

Figure 3-2. A single-player game (you need to fold this sorry hand)

After the "river card" is played, the game determines the winner and all the chips go toward the winning player. The winner's cards are outlined as well, so its easy to spot who has won without comparing everyone's hands (not to mention trying to remember which hand wins: a straight or a flush).

Play Online

The most innovative feature of Texas Hold 'Em King 2 is the online play. You can join public tables or set up your own table that is open to all, or you can require a password to join. You can even maintain a buddy list of players you've played with before to see when they are online.

To join an online game, you'll need to create an account with Magmic, which you can do from within the game. Also, you don't have to register (and therefore pay) to play online—you have full access to all the features during your trial. To create your account, you need to select a username and supply a password and email address. Once your account is created, you're all set. To play online, you will need to have accumulated some money playing the single player mode or the online mode. When you join an online table, you choose what portion of the money you've earned you'd like to bring to the table. Be careful not to join a table with too little money to be competitive. If you have the least amount of money of anyone at the table, you are a prime target. Also, just like in a real casino, you can't go get more chips in the middle of a hand.

If you're still in the middle of a single-player game and you'd like to leave the table and join an online game, you'll be asked if you want to leave the table and take your money so you'll have some to play with online.

The online tables will have real BlackBerry users and AI opponents. The human opponents will have their Magmic username above their head at the table; the AI opponents will look exactly as they do in the single-player mode (see Figure 3-3).

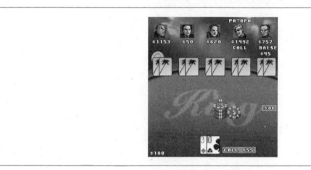

Figure 3-3. An online game with a hand I'm getting ready to fold

If you're not brave enough to join the online games, you can upload the scores you've achieved in the single-player game and see where you rank among other players. Magmic also holds occasional contests with actual prizes for the winners.

Play Chess

HACK #33

Practice to become a grandmaster from anywhere using this third-party game. Take your skills online and compete with other players.

Another winner from Magmic Games (*http://www.magmic.com*) is Medieval Kings Chess 2. The graphics and quality of the Magmic games are impressive, and this one is no different. Medieval Kings Chess 2 lets you play chess on your BlackBerry from anywhere. Once you've honed your skills, you can take your game online to compete against other BlackBerry users.

The program can be installed from Magmic's download page—just follow the same installation instructions as *Texas Hold 'Em King 2* [Hack #32].

Start Playing

Like Magmic's Texas Hold 'Em King 2, this program comes with a seven-day trial. After the seven days have elapsed, you can purchase a key at Magmic's web site for $9.95. To play an offline game against the AI that's built into the software, choose New Game from the main menu. The Create New Game window appears (see Figure 3-4), and you can choose from various settings for your game. You can select the game mode (Player vs. CPU or Player vs. Player), piece color, your opponent's skill level, and the piece set and board type.

Figure 3-4. Create New Game screen

The options for the difficulty of the game range from Beginner (King Arthur) to Very Hard (Will the Conqueror) all the way to Master (Genghis Kahn). The Beginner level, as you'd expect, is quite easy to beat—although it didn't take long for the Master level to dash my hopes of glory!

Use your trackwheel to scroll between settings and the Enter key (or click the trackwheel) to toggle between the values for a particular setting. When you're ready to play, scroll down to Begin Game and plan your attack!

Once the game begins, your pieces will be on the bottom and your opponents will be on the top. The clever controls make this game surprisingly simple to move your pieces. When it's your turn to move, use the trackwheel to scroll between your pieces and use the Enter key (or click the trackwheel) to select one to move. The piece will be highlighted, indicating it is selected. When a piece is selected, the trackwheel is then used to scroll between all the possible moves it is able to make, given the position of the other pieces on the board. In Figure 3-5, I've selected the knight in position B1 and am using the trackwheel to toggle between the two possible moves.

Figure 3-5. The knight in B1 is selected

Online Play

Magmic has built some excellent online features into the game. You can play a random online opponent or set up a private game with a password so you can coordinate a game with a friend. You can also upload your record against the built-in AI to Magmic's online leader boards to compare against other players (*http://www.magmic.com/commons.php?direct=highscores*).

Figure 3-6 shows the online menu you'll see after you log in with your Magmic account and password. You can create an account in the game or on Magmic's web site (*http://www.magmic.com*). You can join and play up to 15 online games simultaneously (I can't imagine anyone complaining that *that* number is too low!). However, the game state is stored on Magmic's game servers, so you can start a game that lasts for days and only make a couple moves a day as your time permits.

Figure 3-6. The online menu

At the time of the writing of this book, the game has only recently been released and the online action is somewhat sporadic, so don't cancel your Xbox Live account just yet. I would expect this game to gain momentum quickly, given the online gaming features and the low price point.

All communication with the Magmic servers flows over HTTP, so your firewall probably won't mind the traffic if you're a BES user using the Mobile Data Service.

HACK #34   Play Free Java Games

There are several free games you can download and play on your device, but don't blame us when your productivity declines.

There are times when the escape that comes from playing a game is just what you need. Whether you're killing time waiting on the next leg of your flight or postponing that project that you've been dreading, there are several free games that will work on your BlackBerry, if you know where to look.

Midlet.org (*http://www.midlet.org*) has a nice repository that has links to the over-the-air installs for several games. Although a good percentage of the games there are no longer available, many still are and will serve your need to procrastinate just fine. Don't expect any game to match the 3D effects and realism of your Xbox, however.

If you're using the BlackBerry Internet Browser, you may have a problem with the *.jad* file formats. Try using the WAP browser (the mMode icon for Cingular, the Download Fun icon on T-Mobile, etc.) or an MDS-enabled BlackBerry Browser if you run into problems.

Monkey Madness

Monkey Madness is available at *http://midlet.org/repository/zendog/monkeymadness/MonkeyMadness.jad*. It is similar to the old Rampage game—you are an 800-pound gorilla that has to destroy buildings and people before they kill you. Use the spacebar to punch the buildings on the screen until they crumble to the ground. There are people in the buildings who are shooting at you, however. It is a race against time as their bullets slowly wear you down (see Figure 3-7).

Figure 3-7. Monkey Madness

Space Travel

Space Travel is a game reminiscent of the old Asteroids game, minus the shooting. It is available at *http://midlet.org/repository/sabbir/spacetravel/ SpaceTravel.jad*. You are a spaceship that has to navigate a space field of asteroids and other objects. You try to collect as many blue moons as you can while avoiding the other objects (see Figure 3-8).

Figure 3-8. Space Travel

Swarm

Swarm is available from *http://midlet.org/repository/zeroindex/swarm/Swarm.jad* and is very similar to the popular Galaga game that you can still find in arcades all over. Use the spacebar to shoot the aliens as they swoop down to try and take you out (see Figure 3-9). Use the S and D keys to move right and left to avoid disaster.

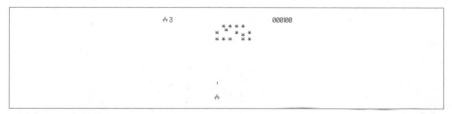

Figure 3-9. Swarm

Spruce Driver

Spruce Driver is part Spy Hunter, part Frogger. It is available from *http:// midlet.org/repository/spruce/driver/SpruceDriver.jad*. The object is to avoid colliding with other vehicles and objects as you race down a highway. Use the S and D keys to move right and left, and try to avoid the puddles—they'll cause you to slip and could cause a collision (see Figure 3-10).

M-Type

Slay a monster using this adaptation of an old game called R-Type. M-Type is available at *http://midlet.org/repository/jshape/mtype/mtype.jad*. Use the

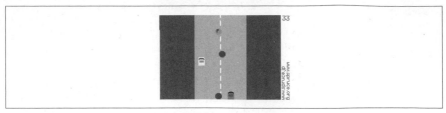

Figure 3-10. Spruce Driver

trackwheel to move your ship up and down and the spacebar to shoot the monster. This one is a little dry, but the graphics are decent compared to some of the other free games (see Figure 3-11).

Figure 3-11. M-Type

Pang

This adaptation of the classic *Pang* is available at *http://midlet.org/repository/ borisgranveaud/pang/Pang_generic.jad*. You try to shoot falling balls before they hit you. When you hit a ball, it breaks into smaller balls. Hit the smaller balls and they finally disappear (see Figure 3-12).

Figure 3-12. Pang

HACK #35 Play Free Magmic Games

Download these free games from BlackBerry's mobile portal—even the original Texas Hold 'Em version.

In an effort to promote the BlackBerry as a worthy mobile gaming platform, Research In Motion has offered a free game from Magmic on its portal. The Bass Assassin fishing game has been available free for some time. Recently, as Magmic Games has become more popular [Hack #32], and they have released more games [Hack #33], RIM has worked out an arrangement to make more Magmic games available for free on its site. At the time of this writing, there are five games available on the BlackBerry portal (*http://mobile.blackberry.com*).

Bass Assassin

Bass Assassin is a game in which you try to catch the most fish in 10 minutes (tournament play) or just fish for the fun of it (see Figure 3-13). To cast, click the trackwheel once. This starts an oscillating meter on the left side of your screen. To make a long cast, click the trackwheel as the red bar is at the top of the meter. To make a short cast, click the trackwheel as it approaches the bottom of the meter. Once you've made a cast, use the trackwheel to speed up or slow down the retrieval of your lure. Try to time it so your lure passes right in front of the fish that you see swimming in the water. Once you hook a fish, your fishing pole bends and a tension meter appears on the left side of your screen. Use the trackwheel to scroll forward to reel in the fish or scroll backward to release tension on the line. If there is too much tension on the line, it will break and you'll lose the fish (and your lure!).

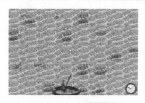

Figure 3-13. Bass Assassin

Raging Rivers

Raging Rivers is a kayaking game in which you try to navigate through a course on a river (see Figure 3-14). You are measured by how long it takes you to maneuver through the white water and obstacles that appear in your path. There are multiple gates that you need to pass through. You'll need to pass through the green gates from the upstream site and the red gates from the downstream side of the gate. If you miss a gate, your time is penalized. Also, watch out for the rocks and the stream's edge—if you touch those too many times, you'll destroy your boat and your chances.

Figure 3-14. Raging Rivers

Spider Solitaire

Spider Solitaire is a variation of solitaire, a single-player card game (see Figure 3-15). The object is to clear all the cards off the table by making an entire row from king to ace of the same suit. Once you complete a row, it is automatically cleared. You can move a card to any column that has the next highest card showing—for example, the jack of hearts could be placed on the queen of spades or the queen of diamonds (or any queen, in fact). To move a card or set of cards, use the trackwheel to place the arrow over the row that contains the card you'd like to move. Select the row by clicking the trackwheel. Scroll to the destination row and click the trackwheel again, and the cards will be moved. You can only move groups of cards of the same suit. The game starts with two decks of cards. Once you run out of moves, scroll to the cards at the bottom right of your screen to deal an extra row of cards.

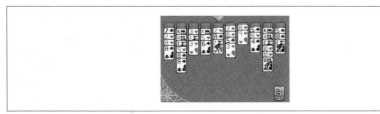

Figure 3-15. Spider Solitaire

Klondike

Klondike is another version of solitaire (see Figure 3-16). The rules are the same as the version of solitaire that comes with Microsoft Windows. You try to move all the cards to the top four slots in order from ace to king. In the lower rows, you can stack cards of opposite color and you can move kings (and all cards stacked on them) to empty rows. Use the trackwheel to scroll among the rows and select the cards you'd like to move. Scroll to the destination and click again to move the cards. Once you run out of moves, scroll to the cards on the top left of the screen to uncover more cards to play. Once you clear the bottom rows of all cards by moving them to the top four slots, you win!

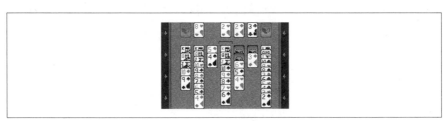

Figure 3-16. Klondike

Texas Hold 'Em King

The first version of the popular Texas Hold 'Em King game [Hack #32] from Magmic is available for free as well (see Figure 3-17). You won't find the online capabilities of the current version in this one, although you can upload your high scores to Magmic to see where you rank in the offline game. This is a great way to learn the game, but be careful—once you start playing this game, you'll want to shell out the $9.95 for the current version! Be sure to bring this one on your next road trip to Vegas!

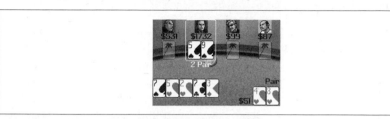

Figure 3-17. The first version of Texas Hold 'Em King

HACK #36 Play Zork on Your BlackBerry

With this Java-based mini Infocom interpreter, you can play retro text adventures on your BlackBerry. You can even write your own games if you want!

"Open mailbox. Take leaflet and read it." Those are the sentences you need to kick off a game of Zork, which is, to many, the prototypical text adventure. You're more likely to hear these games referred to as *interactive fiction* these days, and although many things have changed since the days when Infocom and Adventure International owned the genre, these games are very much alive.

You'll find a thriving community of interactive fiction gamers and authors at the Interactive Fiction Archive (*http://www.ifarchive.org/*). On this site, you'll find hundreds of games, tools to create your own, and all sorts of information.

As a text-friendly handheld, the BlackBerry is the perfect platform for interactive fiction. There are a couple of ways you can play these games on your BlackBerry. Z2ME (*http://gpf.dcemu.co.uk/*) is a port of ZPlet (*http://sourceforge.net/projects/zplet/*), a Java-based interpreter for the Z-Machine, the virtual machine that Infocom used for their interactive fiction games (compiled games are said to be in the Z-code format and have a file extension indicating which version they were compiled for, such as .z3, .z5, and so on). Many modern interactive fiction titles use the Z-Machine, so you'll find plenty of games for it. Another J2ME Z-Machine interpreter is ZeeME (*http://www.gizmo-a-gogo.org/ZeeME/index.html*).

Both Z2ME and ZeeME will run fine on the BlackBerry, but there is a customized version of Z2ME that is more BlackBerry friendly. However, I was unable to get it to work on the 7100 series of BlackBerry handhelds, since Z2ME did not work correctly with the SureType input methods.

Find Games to Play

One of the problems with both Z2ME and ZeeME is that they come with one game (Mini-Zork). To add another game, you'll need to modify the jar file and install it on your device.

Because there is a limit on the size of a jar file (the biggest jar I've been able to load is 76K), you need to pick small games. To see a directory listing of all the Z-code games in the interactive fiction library, visit *ftp://ifarchive.org/if-archive/ games/zcode/*. If you open that URL up in a filesystem viewer, such as the Mac OS X Finder or Windows Explorer, you'll be able to sort by file size. Look for small Z-code files, and cross-reference what you find with the listings at Baf's Guide to the Interactive Fiction Archive: visit *http://www.wurb.com/if/platform*, scroll down to "Z-code," and click the link. Then you can look up a game you're interested in and read a synopsis and review. Not all games are listed, but Baf's is authoritative enough that you safely use the absence of a review as a filter. If you've got all the time in the world, though, you should try everything. There are worse things you could do with your time.

For example, suppose *905.z5* catches your eye. It's 60.5K and might be small enough. So you head on over to Baf's, locate a review, and see that it got a decent review (four stars). Then you download the file, and use the instructions listed in the next sections to package it. Next, create a *.jar* (or *.alx*), post it on a web site for an OTA download (or use the Application Loader) **[Hack #97]**, and start playing the game on your BlackBerry.

> If you do want to post *.jad* and *.jar* files on your own web site, make sure you have the proper MIME types set up **[Hack #97]**.

Use Z2ME

Z2ME is available from *http://gpf.dcemu.co.uk/* as an over-the-air installation (OTA). There are also Application Loader versions available if you have problems with the OTA install (see Phillip Bogle's post at *http://www.thebogles.com/ blog/2005/06/play-infocom-classics-on-your.html* if you are determined to do an OTA install). When you launch Z2ME, Mini-Zork will appear. Figure 3-18 shows an intrepid adventurer doing something very wise.

Figure 3-18. No one wants to be eaten by a grue

Although Mini-Zork will keep you busy for a while, you might want to play some other games. It takes a bit of effort to set this up (and it's lot easier with ZeeME), but it's well worth it. Unfortunately, Z2ME only supports Inform Version 3, and you may not be able to find many compiled games in that format (Versions 5 and higher are more common). But if you search the archive (and Baf's guide), you'll find some.

Suppose you download *dejavu.z3* from the Interactive Fiction archive. You now need to rebuild the Z2ME jar file to hold this new game.

You'll need to grab the source code to the BlackBerry version of Z2ME from *http://gpf.dcemu.co.uk/* and open it in the BlackBerry JDE. Remove *minizork.z3* from the project and replace it with *dejavu.z3*. Then, find the *blackberryZ2ME.jad* file in the project and change the Zfile: entry from:

```
Zfile: minizork.z3
```

to:

```
Zfile: dejavu.z3
```

Save and compile the project, and you'll have a jar that contains a different game.

ZeeME

ZeeME's user interface is a bit rougher than Z2ME's, but it supports newer Z-code files, and also has a nifty packager that lets you easily create jars of games. To use ZeeME on a BlackBerry, scroll all the way to the bottom of the screen, type your command in, and click the trackwheel to bring up the menu. Click Input, as shown in Figure 3-19, to enter your command. Be sure to use the Settings menu to set a comfortable screen width as well.

You can download ZeeMe at *http://www.gizmo-a-gogo.org/ZeeME/index.html*. Both the *.jar* and *.jad* files are included on the site. Since the jar's MANIFEST contains everything ZeeME needs to know to function, you can install directly from the *.jar* rather than from the *.jad*.

As with Z2ME, ZeeME includes Mini-Zork. If you want to run additional games, you'll need to create a separate *.jar* file for each of them. Also available

Figure 3-19. ZeeME's minimalist user interface

on the ZeeME web site is the ZeeME Packager, which is a simple application you can use to create ZeeME jars (you should be able to double-click on the packager *.jar* file to run it). Simply specify the Z-code file, the game name, and the name of the jar you want to create, as shown in Figure 3-20.

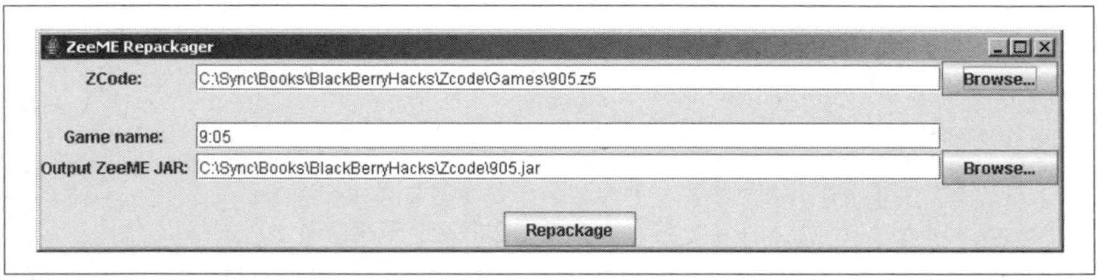

Figure 3-20. Packaging up 905

Post this *.jar* online for an OTA download (as described in "Find Games to Play"), or use the Application Loader to install it. If you need to create a *.jad* for it, it's pretty easy. Use jar tvf *jarfile* (for example, jar tvf 905.jar) to find out the name of the Z-code file (for example, *zeeme/games/905.z5*) and its size. You'll also need to know the size of the *.jar* itself.

Set up the *.jad* as follows:

```
GameTitle0: 9:05
MIDlet-Jar-Size: 73066
GameFile0: /zeeme/games/905.z5
MIDlet-1: Minizork,,zmachine.ui.ZeeMEMidlet
MIDlet-Jar-URL: 905.jar
MIDlet-Version: 0.1.0
MIDlet-Name: 9:05
MIDlet-Vendor: Craig Setera
GameLength0: 61952
```

You can also peek in the jar's *META-INF/MANIFEST.MF* file to get the needed values.

Hack the Hack

It's amazingly easy to make your own game. Graham Nelson's Inform (*http://www.inform-fiction.org/*) will let you create Z-code games. Here's the code

for a simple game where you are trapped in a lousy place until you can get
your BlackBerry working again:

```
Constant Story "My BlackBerry";
Constant Headline
  "^ A simple amusement^";

Include "Parser";
Include "VerbLib";
Include "Grammar";

Object battery "battery"
  with description "A Battery",
  name "battery",
  before [;
    Insert:
      if (second == blackberry && blackberry has open) {
        deadflag = 2; ! indicates you've won the game
      }
  ],
  has concealed;

Object blackberry "BlackBerry"
  with name "blackberry", article "your",
  description [;
    if (self has open) {
      "This is a BlackBerry 7100t; it's missing a battery.";
    } else {
      "This is a BlackBerry 7100t";
    }
  ],
  before [;
    Open:
        if (self hasnt open) {
          give self open;
          "You notice that the battery compartment is empty!";
        } else {
          "That's already open.";
        }
  ],
  has container openable;

Object Startroom "Street"
  with description
    "This is a dark and damp street, and it doesn't feel safe.
    You're standing under the only streetlamp you can see
    for blocks. If you called someone, they'd come and
    pick you up.",
  n_to "That way doesn't feel safe.",
  s_to "Maybe if it wasn't so dark to the south.",
  w_to "It's dark that way, and you hear hungry squishing sounds.",
  e_to "You feel safer under the lamp.",
  has light;
```

```
Object rock "rock" Startroom
  with description "This is a damp and yucky rock.",
  name "rock",
  before [;
  LookUnder:
     if (battery has concealed) { ! It's concealed until we find it
        move battery to player;
        give battery ~ concealed;
        "You found a battery!";
     } else {
        "There's nothing else under the rock.";
     }
  ];

[ Initialise;
  location = Startroom;
  move blackberry to player;
  "^^^BlackBerry Adventure^";
];
```

Download an Inform compiler and the libraries (a bunch of *.h* files) and copy them into the same directory. At the time of this writing, the latest Inform compiler (6.3) was creating Z-code files that were too large to create a reasonably sized *.jar*, so I suggest that you grab a Version 5 compiler and libraries from *http://www.ifarchive.org/indexes/if-archiveXinfocomXcompilersXinform5.html*.

Save the code into a file in the same directory, and call it *bbgame.inf*. When you look at the contents of your working directory, if you see these files, you know you're good to go (if you're on Unix, Linux, or Mac OS X, Inform won't have an *.exe* extension):

bbgame.inf
grammar.h
inform.exe
parser.h
verblib.h

Compile *bbgame.inf* to a *.z5* file with:

```
inform bbgame.inf bbgame.z5
```

You should now have a file called *bbgame.z5*. Use the instructions from the previous sections to package it up in a *.jar* file, and play it on your Black-Berry. For more information about hacking your own interactive fiction games, see *http://www.inform-fiction.org/index.html*.

—*Brian Jepson*

The Internet and Other Networks

Hacks 37–50

With the advent of the Mobile Data Service and TCP/IP on the BlackBerry [Hack #37], an entire world of Internet services became accessible on the Black-Berry. Some of the best applications in existence are accessed by using some type of client software (perhaps just a browser), but the real power comes with its integration with a central service where users and data meet in interesting and exciting ways (think Amazon). Very few client-only software packages carry the same importance as one that integrates well with a web service. With your BlackBerry, you can track your to-dos [Hack #47], corral your bookmarks [Hack #46], and even use instant messaging [Hack #44].

 HACK #37 ## Configure Your BlackBerry for Internet Access

Unlock the PC in your BlackBerry by giving it a TCP/IP stack.

When you access your BlackBerry Browser, you are communicating in a limited way through a gateway of sorts. Your BlackBerry doesn't need its own IP address to communicate in this fashion. You can, however, depending on your provider, enable TCP/IP on your device. This allows your device to communicate directly (well, almost directly) with machines on the Internet.

Enabling TCP/IP on your device gives you access to many of the network-enabled third-party applications for the BlackBerry, some of which are highlighted in this book. If you are not a BlackBerry Enterprise Server user, this hack will give you many of the functions of the BES and MDS without being BES enabled.

Enable TCP/IP on Your Device

This hack requires that you have BlackBerry 4.0 or later installed on your device. If you have a version earlier than 4.0, you should install the latest version on your device [Hack #20]. Starting with 4.0, there is an additional item in your Options application called TCP, as shown in Figure 4-1.

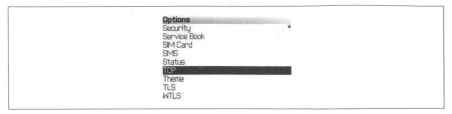

Figure 4-1. The TCP option in the Options program

Inside the TCP settings, you'll find three fields that you will need to configure. These settings will be specific to your provider. The APN stands for *Access Point Name*. Every provider requires a value for the Access Point Name field. Some providers require that you fill in the username and password while others allow you to use null values for these fields. A good resource to find these settings is the Opera Access Point settings page available at *http://www.opera.com/products/mobile/docs/connect/*. Figure 4-2 shows the settings required for Cingular.

Figure 4-2. The Access Point settings for the Cingular network

What Does This Do?

You won't notice anything different about your device immediately upon enabling this. However, you have opened up a whole new world of network-enabled applications to use on your device! You can install and use different browsers [Hack #66] and email programs [Hack #67], as well as instant messaging programs [Hack #44] that integrate nicely with your BlackBerry alerting; shop on Amazon.com [Hack #58]; and even remote control your desktop machine with VNC [Hack #40]!

H A C K   Blog from Your BlackBerry
#38   Why should you blog only from your desktop computer?

You have just thought of a brilliant idea that you have to blog about right away, so like a caveman, you scribble it down on a piece of paper and hope you don't lose it before you return to your computer. You might not even remember to blog when you return to your computer, and your precious idea is already turning stale anyway.

It doesn't have to be this way. With most blogging software, you can turn on email-to-blog functionality so you can blog from any email-enabled device—a domain that is ruled [Hack #31] by your BlackBerry.

Turn On Email-to-Blog Functionality in Blogger

From within Blogger, Google's popular blogging service, log in and go to your blog administration page. Click on the Settings tab, and then click on the Email heading and go to the Email-to-Blogger section.

Blogger has you create a "secret" email address that allows any messages that get sent to it to be automatically blogged. As shown in Figure 4-3, the email address is constructed with your Blogger name, followed by a string of your choosing. The Publish checkbox tells Blogger whether to go ahead and publish the content of messages as they are sent or to add it to your Blogger admin page without publishing, giving you the opportunity for review before your blog visitors see it.

Mail-to-Blogger Address   davemabe.`thisisasecret`   @blogger.com ☐ Publish

This is an address by which you can post to your blog via email. The secret name must be at least 4 characters long.

Save Settings

Figure 4-3. Blogger's email-to-blog settings

Write by Email in Wordpress

Wordpress is another popular blogging software package that is written in PHP and runs on Apache and has a MySQL database backend. Along with a mountain of excellent plug-ins, it has a built-in feature that allows you to publish to your blog by email.

Log into your Wordpress administration page and click on Options → Writing. This brings you to the Writing Options page as shown in Figure 4-4.

This Wordpress functionality requires a little more work than Blogger's does. You have to have already created an email account that is accessible via POP3 (a free, ad-hoc Gmail account is a good choice because of its built-in spam protection). You configure the POP3 settings to tell Wordpress where to retrieve messages from. Use the "Default post by mail category"

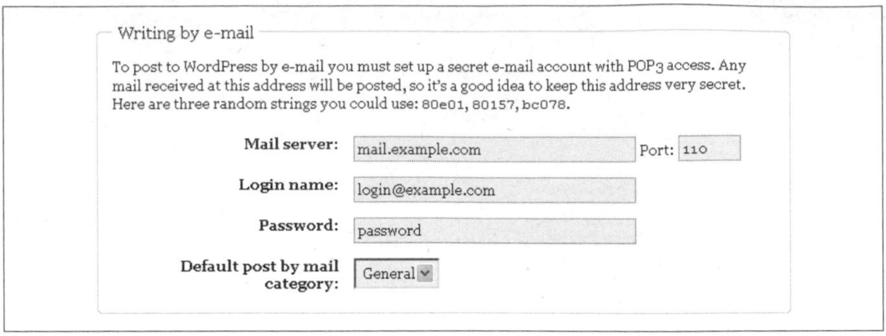

Writing by e-mail

To post to WordPress by e-mail you must set up a secret e-mail account with POP3 access. Any mail received at this address will be posted, so it's a good idea to keep this address very secret. Here are three random strings you could use: 80e01, 80157, bc078.

Mail server: `mail.example.com`   Port: `110`

Login name: `login@example.com`

Password: `password`

Default post by mail category: `General`

Figure 4-4. Wordpress writing-by-email option

option to choose one of your categories that all posts by email will go into. Of course, you can always go back and change the category and even add additional categories after a post has been published.

Once you've sent a message to the blog-by-email account you've set up, you'll need to visit this page: *http://your.blog.url/wp-mail.php* (*wp-mail.php* is in the same directory as your main *index.php*, so the URL could vary depending on your setup). This scans your mail and posts any messages in your inbox to your blog.

Many Wordpress users choose to automate this step by using a scheduled job to make the request to *http://your. blog.url/wp-mail.php* on a regular basis. There are a variety of tools you can use to make HTTP requests. One tool is *lwp-request*, and it comes with ActiveState Perl, a tool that is used in several of the hacks in this book. (If it didn't come with your Perl intallation, you can obtain *lwp-request* as part of the libwww-perl collection at *http://lwp.linpro.no/lwp*.)

Blog Away

Once you've configured your blogging software to accept email, go ahead and compose a message from your BlackBerry. The subject line will become the title of your post, and the body of the message will become the post content. Figure 4-5 shows a composed email message I've sent to my Blogger blog.

To: My Blogger Blog
Subject: A Great Idea

I had a great idea and I just have to blog about it before I forget

---

Figure 4-5. Blogging from a BlackBerry

Almost instantaneously, the blog appears on my blog, as if I posted it from my desktop browser (see Figure 4-6).

Figure 4-6. My email-to-blog entry

 You don't want spam finding its way onto your blog (what a nightmare that would be!), so you need to choose an email address that it not easily guessed and is longer than a few characters. Blogger enforces a length of at least four characters, but I would highly recommend one that is much longer. You'll only have to type it once into your address book, and then you'll just pick it from your list when you compose a message, so make it complex and lengthy, just as you should for an important password.

See Also

- Bloxsom (*http://www.blosxom.com/*)
- Nucleus (*http://dev.nucleuscms.org/*)
- MovableType (*http://www.sixapart.com/movabletype/*)

 HACK #39

Read News and Blogs on the BlackBerry

RSS and Atom feeds offer a smaller version of a web site, but some or all of the same content. By using an aggregator, you can cluster all sorts of related sites together and read news and other data from a variety of sources large and small.

There are a variety of ways to get your RSS feeds onto your BlackBerry, but my favorite is to use Bloglines (*http://www.bloglines.com/*).

Bloglines is a web-based RSS reader, which sounds kind of goofy in a way because you're trying to leave the web browser behind when you're scouring over your feeds. Fact is, Bloglines is quite brilliant due to one critical reason: it keeps track of what you're reading no matter where you're logged in from. When you log in via your desktop PC, your PowerBook, a computer at a friend's, a public terminal, or your BlackBerry, you'll never see the same article twice if you use Bloglines (unless you mark it "Keep New"). Using Bloglines will cut down dramatically on the flood of things you are confronted with at any given time, and certainly helps make your RSS feeds more useful on the go.

Using Bloglines via its default web interface isn't the best way to view it on your BlackBerry, however. You may like using Bloglines Mobile (*http://bloglines.com/mobile/*) in the BlackBerry browser, which will work fine.

An alternative is using Berry Bloglines (*http://www.thebogles.com/berry_bloglines.html*), a fantastic utility that connects you to Bloglines via a little proxy server that converts news items into easier to read formats for the BlackBerry. It optimizes the content it receives via some magic with Python. Although Phillip Bogle did not toot his own horn about this application in this book, I'll do it for him: Berry Bloglines is *awesome*.

There is a possible alternative to Bloglines Mobile called litefeeds, which is still undergoing development (see *http://www.litefeeds.com/*). They have a Black-Berry client application in the works, and, last I checked, it was a service with a lot of promise. Keep an eye on it. Litefeeds is specifically geared for mobile RSS, and it is always possible that Ask Jeeves (the current owner of Bloglines) will make Bloglines lame in some way, so it'll be very nice to have a service to step up should the worst happen. Best case scenario, Bloglines isn't the only game in town anymore and the two of them can keep each other on their toes.

If you prefer the traditional way of reading RSS feeds using a dedicated client application, try Newsclip from Virtual Reach (*http://www.virtualreach.com/*) and Berryvine RSS (*http://www.berryvine.com/*). In my opinion, they're the best of the breed for dedicated RSS feed–reading on the BlackBerry. The other applications in this category just don't even come close to it. Both Newsclip and Berryvine's RSS have trial versions available, letting you examine all their features without any cost to you.

Newsclip
> Virtual Reach includes a nice index of possible RSS feeds for Newsclip users, much like NetNewsWire or other desktop RSS aggregators. You can also add your own RSS feeds to subscribe to, and read them on the go. What tends to work best for me is to put only the critical RSS feeds on my BlackBerry in Newsclip. This usually means a couple of my favorite Craigslist RSS feeds and some work-related feeds that let me feel like less of a slacker while in line at the DMV.

Berryvine
> You can differentiate feeds in Berryvine RSS by applying colors to them— green for General, orange for Entertainment, etc. You can also categorize feeds and view feeds by category, which is a helpful trick as well.

It's quite easy to get your feeds on a BlackBerry. It is one of the areas where a lot of development hours are being burned, which is fortunate for all of us BlackBerry users!

—*R. Emory Lundberg*

Control Another Computer Remotely

#40 Use this open source remote control software to access other machines from your device.

Wouldn't it be nice to access the desktops of other computers when you only have access to your BlackBerry? It turns out you can by using Virtual Network Computing software, or VNC. VNC allows you to remotely control another computer across a network. You have access to the keyboard and mouse, and you can even send a Ctrl-Alt-Delete to a Windows computer to shut it down. For Windows and Linux, you should check out the free TightVNC program (*http://www.tightvnc.com*). It is compatible with most VNC clients, but will optimize bandwidth when used with TightVNC-compatible clients. Mac users can use the free OSXvnc (*http://www.redstonesoftware.com/vnc.html*). There is VNC support in recent versions of Mac OS X (check out Apple Remote Desktop in System Preferences → Sharing).

There are ports of the client and server components of VNC for practically any operating system in existence and, because it's open source software, its features and stability continue to improve.

Install the Server

To use VNC, you have to install the server software on the machine you'd like to remote control. You can download and install the server software from the VNC web site at *http://www.realvnc.com*. Choose the appropriate software for the operating system you'd like to remotely control and follow the installation instructions.

Install the BlackBerry VNC Client

The VNC client for the BlackBerry device is available at *http://www.ethell.com/j2mevnc/*. There is no over-the-air install available, so you will have to download the archive and use Application Loader to install the software on your device. The latest version can be downloaded using this URL: *http://www.ethell.com/j2mevnc/VNC-en-bb.zip*.

Once installed, there will (oddly enough) be two icons for VNC on your Home screen, as shown in Figure 4-7. One icon runs the actual program and the other displays an About screen describing the program. So, this would be a good opportunity to hide the About VNC icon from the Home screen **[Hack #5]**.

Figure 4-7. Two VNC icons on the Home screen

Connect to a VNC Server

When you run the program from the Home screen, you'll be taken to the VNC client that will allow you to connect to a remote computer running the VNC server component. To make a connection, enter the name or IP address of the computer to which you'd like to connect in the Host field. If the computer requires a password (and it should!), enter it in the Password field. Figure 4-8 shows the screen you'll see when making a connection.

Figure 4-8. Connecting to a remote computer

The Share Desktop option should be enabled if you'd like other VNC clients to be able to connect to your session on the VNC server at the same time you are connected. The NCM option stands for *Nokia Compatibility Mode* and can be ignored.

Use the trackwheel to bring up the menu and click Connect. The client then attempts to make a connection to the VNC server, as shown in Figure 4-9.

Figure 4-9. Connection attempt

Once connected, you'll be able to view the desktop and use the mouse on the remote computer. To use the mouse, click the trackwheel and select Mode. Under the Game Mode section (see Figure 4-10), enable the Mouse Mode option.

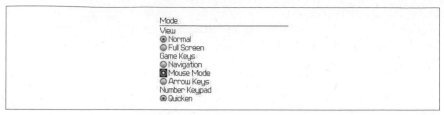

Figure 4-10. The Mode config screen

A mouse pointer will appear in the screen, as shown in Figure 4-11. If you're used to using a standard desktop version of the VNC client, you'll notice the familiar red dot that the remote mouse pointer follows around the screen.

Figure 4-11. Remotely controlling a VNC server

I'm not exactly sure why, but there is a full screen mode included for those of you who have the vision of a hawk. I wouldn't recommend trying to use it.

 Because you are operating over a GPRS or other cellular-based network, you will notice a significant delay using the client. If you are used to the desktop version of VNC, you'll probably grow weary waiting for the screen to refresh on your BlackBerry. Of course, when you are in a bind, this could come in handy.

If you're willing to spend a little money, there is an excellent remote control client and server that works well on the BlackBerry called Remote Desktop for Mobiles (*http://www.desktopmobiles.com/*). It has better performance than VNC, although it works only on the Windows platform.

HACK #41 Use Google Maps

Although Google provides access to maps for handhelds through Google Local, this handy program works very similarly to the full browser implementation of Google Maps.

It's hard to overestimate the impact on the online map industry that Google Maps (*http://maps.google.com*) has had. The crisp, readable maps, along with the integration with Google Local, have made quite a splash in the online community. But most agree that the most innovative feature of the Google Maps interface is its clever use of JavaScript to create a user experience in a simple web browser, rivaling the most polished of desktop software without requiring browser plug-ins.

Use Mobile GMaps

You can use a program called Mobile GMaps (*http://www.mgmaps.com/*) to access Google Maps from your BlackBerry. It is available for free and you can download it over the air right from your device by going to their WAP interface (*http://wap.mgmaps.com/*).

Once installed, use the MGMaps icon on the Home screen to open the program. You're greeted with a map of the United States (see Figure 4-12), the same default view you'll find on Google Maps (*http://maps.google.com*).

Figure 4-12. The default view in Mobile GMaps

Use the trackwheel to access the menu and select Search to enter an address to search for. You can perform searches, such as restaurants in New York, NY, or you can search for a specific address in the Where field. You can also choose whether to view a regular street map or a satellite view. Once your search results are returned, use the trackwheel to highlight one of the results and choose Select from the trackwheel menu. This will download the appropriate images from Google Maps to your BlackBerry and display them as shown in Figure 4-13.

Use GMapViewer

Google Maps's sophisticated use of JavaScript (a technique known as AJAX) is not yet available in any handheld browser, but there is a free program for handhelds called GMapViewer (*http://www.sreid.org/GMapViewer/*) that creates a look and feel that is similar to what the Google Maps interface provides to desktop browsers. Because it is written according to the J2ME specifications, you can run it on your BlackBerry!

Figure 4-13. W Main Street in Carrboro, NC

Go to the GMapViewer web site (*http://www.sreid.org/GMapViewer/*) and use the over-the-air download to install the application. After you install the program, click on its icon on the Home screen to execute it. Figure 4-14 shows the very simple screen that appears when you bring up GMapViewer.

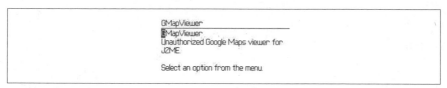

Figure 4-14. GMapViewer Home screen

View a Map

To view a map, click the trackwheel from the menu and select the Search menu option. This brings you to a screen to create searches for cities and addresses, as shown in Figure 4-15. Select New search from the menu and enter an address or city for a place for which you'd like to view a map. You'll need to enter the address all on the same line.

Figure 4-15. Choosing New search from the menu

Click the trackwheel and choose OK. The search you entered is sent to the service's web interface (more on this later) and is validated. Once the search returns, use the trackwheel to choose Select from the menu, and you will be taken to the map for that location in another screen.

Navigate the Map

You will notice a similar look and feel to the click and drag interface you're used to in your desktop browser. Use the trackwheel to scroll to the edges of the map, and the new sections will be filled in on the fly. Figure 4-16 shows the program as you scroll to the map's edge. Notice the orange and blue blocks that appear for areas on the map that have yet to be downloaded. Also, the wireless activity arrows indicate that the missing images are being retrieved onto your device.

Figure 4-16. Scrolling to the map's edge

Add a Map Pin

You can add new map pins, which will be stored in a menu for easy access. There is a crosshair in the middle of the map screen that stays centered as you scroll around the map. At any time if you would like to create a *waypoint* as a location bookmark, click on the trackwheel and choose New Map Pin from the menu. This brings up a text entry field in which to enter a string that identifies the location. After you've typed the text for your map pin, choose OK from the trackwheel menu. You'll be returned to the map you were viewing, and you'll see a red map pin with the text you entered on the map, as shown in Figure 4-17. The map pins you create are stored on your device. You can go to the Map pin menu to view all the pins you've created and go directly to that section on the map.

Figure 4-17. A custom map pin

How Does This Work?

GMapViewer is used in conjunction with a web service. This web service acts as a gateway and runs as a PHP application that can be downloaded along with the source code for the J2ME client software. By default, the client points to a copy of the web service running at the author's web site. Be sure and check the GMapViewer web site for updates—the author currently states that the gateway could be taken down at any time.

If the gateway was put out of service, I would anticipate another kind soul to provide a different gateway that clients could point to. Because both the client and gateway components are released under the *GPL* (GNU Public License), you could download the source code and run a GMapViewer gateway on your own web server!

HACK
#42
Telnet or SSH to Internet Servers

Tail that logfile from anywhere you have cell coverage with the MidpSSH telnet/SSH client.

Once you've configured your device for TCP/IP, you can use open source Telnet and SSH clients to connect to a server on the Internet. No doubt you won't be able to use your god-like *vi* skills on your BlackBerry keyboard, but this will give you the full SSH access you sometimes can't live without.

Install and Use MidpSSH

Similar to most applications, MidpSSH (*http://www.xk72.com/midpssh/*) provides over-the-air installs and downloads for installing via Desktop Manager and Application Loader. MidpSSH also provides various versions of their software. You can get the full version that has support for any of the various protocols you'd need to use: Telnet, SSH1, SSH2, and various combinations of these. The space-conscious user will like the ability to install a version of the software that contains support only for the protocols they need and nothing more.

Be sure to choose the BlackBerry-specific version of MidpSSH. The generic SSH1 and SSH2 builds (full and lite) will probably not work on a BlackBerry.

Once installed, select Sessions and use the trackwheel to choose Select from the menu. A screen appears, allowing you to define the settings for your connection, as shown in Figure 4-18. Scroll down to enter your username and password. Click the trackwheel once and choose Create from the menu to save your new connection settings.

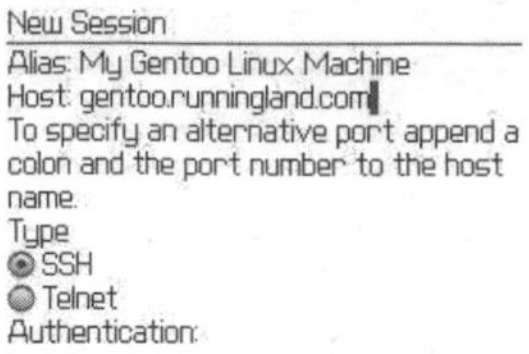

Figure 4-18. Defining the session properties

After you've created a new session, use the trackwheel to bring up the menu and choose Connect, as shown in Figure 4-19.

Figure 4-19. Making the connection

Enter Commands

MidpSSH connects to the remote computer and uses the username and password you defined in the connection to authenticate so you don't have to type it (thank goodness!). Although it appears you should be able to start typing commands just as you would from a computer, MidpSSH forces you to use the trackwheel to choose Input from the menu. It gives you a blank screen in which to type your full command and, unlike the SSH client you're probably used to, the characters you type aren't immediately sent to the remote machine. MidpSSH allows you to type your command and then send it all at once to the remote computer. This gives you a little nicer interface for typing and allows you to correct the mistakes you'll inevitably make before sending the command. Figures 4-20 and 4-21 illustrate the process of entering a command in MidpSSH.

Figure 4-20. The input menu on a MidpSSH connection

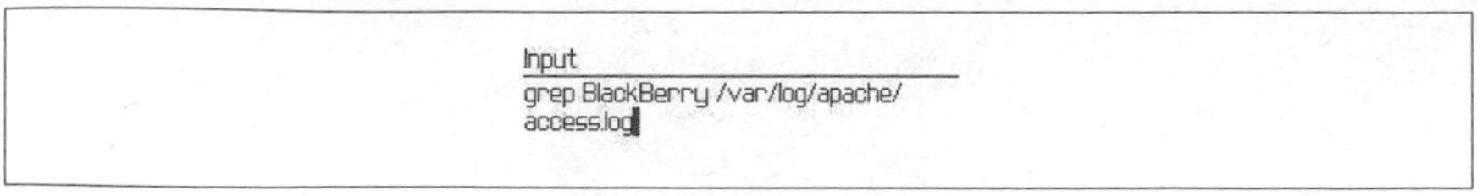

```
Input
grep BlackBerry /var/log/apache/
access.log
```

Figure 4-21. Type (and correct!) your command before sending

The default font size for your connection is very tiny. You will probably want to change it to something a little larger. Go to Settings → Font to access that setting. Of course, as you increase the font size, you'll lose valuable screen real estate. I found a setting of *small* to be the optimal setting for my use.

More Info

Although you can access Internet servers with MidpSSH, you won't be able to reach your intranet computers (unless, of course, they're accessible from the Internet). However, the price is right (free!) for this software. If you need access to intranet servers, use the Idokorro client [Hack #43].

Telnet or SSH to Intranet Servers

#43 Use the Idokorro client to Telnet/SSH to intranet servers through your BES's Mobile Data Service.

The MidpSSH client [Hack #42] can be used to connect to computers on the Internet, but what about connecting to servers on your intranet? There is a client from Idokorro (*http://www.idokorro.com/*) called Idokorro Mobile SSH that can do just this. You must be a BES user to use this client on your intranet, since it uses the BlackBerry Enterprise Server's Mobile Data Service (BES/MDS) to connect. If you're a BlackBerry Web Client only user, you can set up TCP/IP [Hack #37] on your device and use this software in TCP-only mode.

Install and Configure Idokorro SSH

Idokorro Mobile SSH is installed the same as most other applications: there is an over-the-air install or a download to install with Application Loader. Once installed, select the icon from your Home screen. The first time the software runs on your device, it will ask you whether you want the program to use your BES/MDS or TCP/IP for its network communication, as in Figure 4-22.

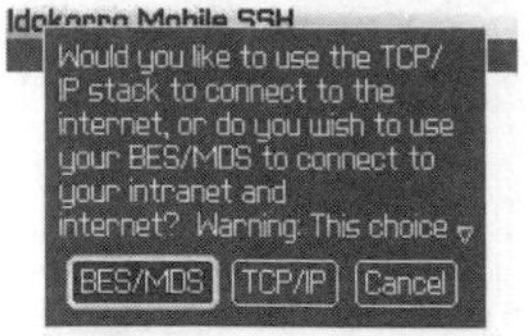

Figure 4-22. Choosing the mode depending on your connection

Notice the stern warning that is given for the connection type. It warns that the choice is permanent and the only way to change from BES/MDS to TCP/IP or vice versa is to reload the entire device. Whoa—that's pretty ominous! Is this just bad software design by Idokorro? Actually, no. The Black-Berry operating system is designed with security in mind and it is able to detect when an application that uses TCP/IP later switches to BES/MDS. When this situation is detected, the operating system disables the third-party application for-ever, or until you reload the entire device. This is to prevent applications from accessing corporate data and then immedi-ately being able to make a connection to the Internet, pre-sumably to steal proprietary data. Most vendors distribute two versions of their software, one solely for BES/MDS users and another for TCP/IP users. Idokorro has chosen not to take this route and, quite frankly, has made it a little less convenient for its users.

Create a New Connection

The connection settings screen for the Idokorro software is similar to the MidpSSH client, although there are important differences. The Idokorro cli-ent lets you control settings that MidpSSH doesn't. You can control line wrapping as well as the number of lines in the buffer for scrolling back. You can also specify your own colors for the foreground and background in your connection. You also have more control over which protocol you'd like to use when you connect. For example, the client lets you force the connection to connect using the SSH2 protocol.

You can use SSH2 with MidpSSH, but there will be quite a delay when the client negotiates the keys with the SSH server. Idokorro Mobile SSH overcomes this liability by using a built-in RIM library to perform the negotiation, which speeds up the key generation considerably.

Like most modern SSH clients for your computer, Idokorro will let you know when the server's key fingerprint is new or updated, as shown in Figure 4-23. This serves as an alert that there could be a security issue or DNS tomfoolery.

Figure 4-23. The server's key fingerprint is not in the local cache

Connect to a Remote Computer

The Idokorro client seems more polished than the MidpSSH client when making connections. For one, it respects some of the usability conventions of the BlackBerry. For example, the client allows you to use the Enter key instead of having to use the trackwheel to access the menu when performing the default function. You'll also notice familiar dialog boxes like Save, Discard, or Cancel when you've made a change without saving. This is a refreshing change from the far too many third-party BlackBerry applications that show no regard for these usability guidelines.

The software gives excellent status and feedback as you attempt to make a connection to a remote machine. There is also a nice label at the bottom of the connection screen reminding you which machine you're currently connected to (see Figure 4-24).

Figure 4-24. Connected to a remote machine

That label becomes your text input area when you type commands. Simply start typing a command and the label disappears and your command appears in its place (see Figure 4-25). Hit the Enter key when you've completed your command to send it to the remote machine. This is far more usable that the MidpSSH client in which you have to click the trackwheel a minimum of twice to enter a single command.

Figure 4-25. Entering a command

Chat over IM

IM is becoming an essential business tool. Use this program to stay logged into several IM clients at once while you're on the go.

IM isn't just for your teenage children anymore! There is no denying that instant messaging has come into the mainstream. In many businesses and industries, it has become as ubiquitous and as essential as email. If you'd like to stay connected on IM on your BlackBerry, you can use a program called VeriChat to connect to various IM services from your device. You may also want to check out the BlackBerry Messenger that RIM is including in the latest handheld software [Hack #20]—it is an instant messaging client that rides on top of PIN messaging [Hack #27]. You can use the BlackBerry Messenger only with other BlackBerry users, whereas you can use the VeriChat client to communicate with IM users on a variety of services and protocols.

Install and Configure VeriChat

There is an over-the-air download of VeriChat available at *http://www.pdaapps.com/ota/VeriChat.jad* or, if you prefer to install using Application Loader, you can download the install from *http://www.verichat.com/verichat_bby/index.html*. Once installed, run the program from the Home screen, and you'll immediately be able to set up your instant messaging accounts for MSN, Yahoo!, AIM, and ICQ. Figure 4-26 shows the configuration of the MSN client. For each protocol you configure, use the trackwheel to bring up the menu and choose Continue to proceed to the setting for the next protocol. If you don't want to configure an account with a certain protocol, choose Skip/Delete from the menu.

Chat with Verichat

After you've configured your accounts, you can log in and begin chatting. Figure 4-27 shows the professional-looking login screen that displays as VeriChat logs you in to the various IM protocols you've set up.

After logging in, your buddy lists are displayed (VeriChat pulls them down from the IM network you've logged into so you don't have to type them all

Figure 4-26. Configuring MSN IM in VeriChat

Figure 4-27. Logging in with Verichat

over again on your BlackBerry). You can control whether all your buddies appear in the list or whether only you're online buddies are displayed. Given the size of the screen, you'll find that displaying only your online buddies will save you valuable screen real estate.

As Figure 4-28 shows, your buddies appear in alphabetical order, regardless of the IM service through which you're connected to them. The icon next to the buddy name indicates which service's buddy list that user is on.

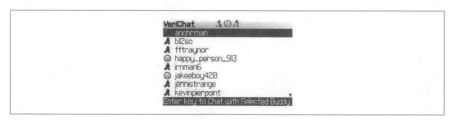

Figure 4-28. Your buddy list

VeriChat integrates nicely with the BlackBerry and conforms to the standards you're accustomed to. For example, most normal chat functions don't require use of the trackwheel—just click the Enter key to start a chat with a particular user. Of course, if you prefer the trackwheel, you can use it for most functions as well.

Once logged in, you can have VeriChat run in the background, and then you can continue your normal, everyday BlackBerry routine. When you get a message from someone, you'll get a notification, as shown in Figure 4-29, no matter what BlackBerry program you happen to be using at the moment.

Figure 4-29. You've received a message

As you would hope, selecting No simply returns you to exactly what you were doing before. Selecting Yes on this screen takes you right into a chat session with your buddy, as shown in Figure 4-30, where you can chat much like you would on a desktop computer.

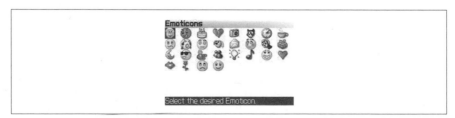

Figure 4-30. Chatting in VeriChat

You have access to *predefined messages* that allow you to send a commonly used response without typing it. Figure 4-31 shows the ones that are available in the program by default. You can add your own by clicking the trackwheel and selecting Add from the menu. You can also use emoticons in your chats (see Figure 4-32). Note, however, that the actual icon won't appear in your chat window—only the text that represents the emoticon.

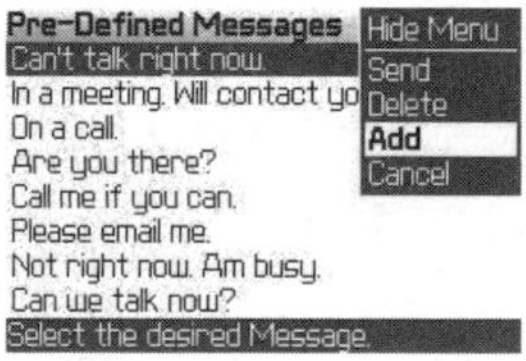

Figure 4-31. Adding a new predefined message

Figure 4-32. Emoticons in VeriChat

See Also

There are a few other IM clients for BlackBerry that are worth looking into:

- Causerie IM for BlackBerry (*http://www.mantragroup.com*)
- IM+ for BlackBerry (*http://www.shapeservices.com/eng/im/BLACKBERRY/*)

HACK #45 Chat over IRC

Enter chat rooms from anywhere by using this free client software on your device.

Ah, IRC. Internet Relay Chat is a great tool to find information or a good place to get blasted for asking stupid questions with a response of "RTFM." The great thing about the world of technology is that when you run into a problem, there is very little chance that you are the first one that has ever experienced it. In fact, there is probably a motivated user of the same software you are having a problem with lurking out on IRC that has the answer and is willing to share. Many high-profile techno-geeks hang out on IRC providing newbies with answers to common problems.

Good news! IRC is no longer just for your desktop computer. There are client versions that work on many J2ME phones—including your BlackBerry. There are a few versions to choose from, but the client that seems to be the most stable and has the most features is WLIrc (short for Wireless IRC), and is available for free from *http://wirelessirc.sourceforge.net/*.

Install WLIrc and Connect

Using your BlackBerry Browser, go to *http://wirelessirc.sourceforge.net/* and click the over-the-air installation. There is also a zip archive for installing with Desktop Manager and Application Loader. Once installed, you'll see a tiny icon on your Home screen, as shown in Figure 4-33.

Figure 4-33. The WLIrc icon on your Home screen

WLIrc supports two types of connections: sockets to connect directly to an IRC server or HTTP to use a proxy to connect to an IRC server on your behalf. It is recommended to use sockets if your device and service supports it, although HTTP seems to work just as well.

There are countless IRC servers available on the Internet for you to connect to. One of the most popular is *irc.freenode.net*. Once you've decided which server to connect to, use the trackwheel to access the menu and click the Configuration option. Enter your chosen IRC server in the "Irc server" field (see Figure 4-34) and choose Save from the menu using the trackwheel. You'll also want to customize your nickname using the Nick field. If the nickname you've chosen is already taken, you'll be prompted to enter a new one as you make your connection.

```
Config
Nick DaveMabe
Real name WLirc user
Channels #WLirc
Irc server irc.freenode.net
Irc server port 6667
Notifylist seperated with ' '
```

Figure 4-34. Entering your IRC server name

After you save your configuration settings, choose Connect from the track-wheel menu to make your connection and log on as shown in Figure 4-35.

```
WLirc                    Hide Menu
WLirc 0.97               Connect
By TheSverre             Configuration
                         Advanced
                         Change font
                         Exit
                         Select
                         Close
```

Figure 4-35. Connecting to the IRC server

After logging on, you'll be connected to your default channel, which you can customize using the Configuration settings mentioned earlier.

Join Channels

Choose Join from the trackwheel menu while connected and enter the channel you'd like to join. This brings you to the Join channel screen where you can type the name of the channel and choose Ok from the menu, as shown in Figure 4-36. There isn't a way to bring up a list of all available channels to choose from directly as there are on desktop IRC clients, so you'll need to know the exact channel name to join. Figure 4-37 shows a connection to the #Perl IRC channel. The number in brackets on the top beside the channel name is the number of users currently connected. You can use the Names option on the trackwheel menu to view that list of users.

Figure 4-36. Joining the #Perl channel

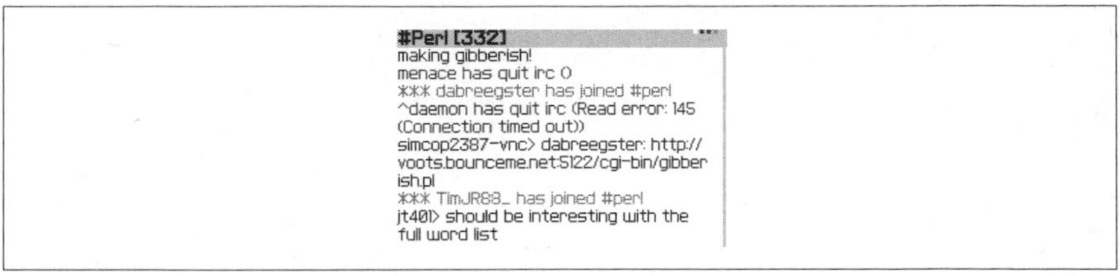

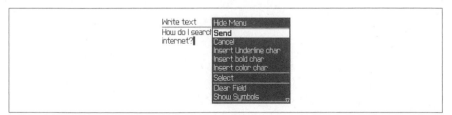

Figure 4-37. The #Perl IRC channel

One feature that sets WLIrc apart from other handheld IRC clients is the ability to stay connected to multiple channels at the same time. You just repeat the same technique to connect to more than one channel at once. After you're connected to more than one channel, you can switch back and forth between them by clicking on the trackwheel and choosing Windows from the menu. This presents you with a list of the channels you're currently connected to along with your current connection status window. Scroll to the channel you like to display on your screen and use the trackwheel to choose Select. The small rectangles in the top-right corner represent the current channels (or windows) you're connected to.

Send Messages

When you are in a channel, you have access to similar functionality as your desktop IRC client. Use the trackwheel to access the menu and select Msg. This puts you in a Write Text screen where you can type the message you'd like to send. When your message is complete, choose Send from the trackwheel menu, as shown in Figure 4-38. Sending messages on WLIrc is much the same as any IRC client.

Figure 4-38. Sending a message in an IRC channel

Bookmark It on del.icio.us

You can use the excellent social bookmarking service to do a variety of things with your BlackBerry.

If you haven't discovered del.icio.us yet, you have to go check it out at *http:// del.icio.us*. It is not just a clever URL. It is a *social bookmark* service that allows you to add bookmarks and assign tags to them. You can then search for every link to which you've assigned a certain tag, or search for any link to which anybody has assigned a certain tag. It also allows you to store bookmarks in a single location that is accessible from any computer that has Internet access. As you'll see, this comes in quite handy with your BlackBerry.

The del.icio.us service has gained a great deal of steam in the past few months. They've released an API that motivated users can exploit to create new ways to use del.icio.us. Expect to see more and more clever ways to use del.icio.us.

Start off by visiting *http://del.icio.us* in your desktop browser and create an account. Once created, familiarize yourself with the interface and posting new links to your bookmarks. There are tiny plug-ins that are available for use with most any modern browser that make posting new links to del.icio.us a breeze.

Post from Your Device

One of the situations I've found most useful is when I am browsing the Web using the BlackBerry Browser and come across a URL that I know I'd like to read, but because the site is naively formatted only for desktop browsers, it makes it nearly impossible to read on the device. When this arises, I post the URL to del.icio.us with a special tag called "desktop." Because del.icio.us allows you to subscribe to RSS feeds for any tag, I've subscribed to my "desktop" tag feed using Bloglines. When I check Bloglines from my desktop computer, I'm automatically reminded that there was an interesting URL that I'd like to read.

To post to del.icio.us from your device, do the following. First, visit a site you'd like to post to del.icio.us. Click on the trackwheel to bring up the menu and select Page Address, as shown in Figure 4-39.

Figure 4-39. Selecting Page Address from the menu

This brings up a dialog that shows the URL and the title of the site. Use your trackwheel to scroll down to the Copy Address option on the dialog to copy the URL to your device's clipboard. Next, go to *http://del.icio.us/post/* and log in if you aren't already (you'll probably want to add this URL to your BlackBerry bookmarks). Scroll to the URL field, click the trackwheel to bring up the menu, and select Paste. Figure 4-40 shows the pasted URL in the field on the del.icio.us posting page.

Figure 4-40. Posting to del.icio.us

Click the save button and, on the next screen, enter a title for the page and scroll down to the tags field. Type "desktop" for the tag (see Figure 4-41), along with any other tags you'd like to assign to the link.

Figure 4-41. Assigning your special tag to the link

Click the save button, and your link will be added to del.icio.us and you will be redirected back to the site you were viewing. Later, when I've returned to my desktop computer, the link that I posted from my device appears as a new item in Bloglines (see Figure 4-42).

Figure 4-42. My new link in Bloglines

Manage Your Device Bookmarks from Your Desktop

Most people have hundreds of bookmarks in their desktop browsers, so you probably would like only a subset of your desktop bookmarks to be available on your BlackBerry. You can always manage them from your device either by manually creating a bookmark in the BlackBerry Browser by typing the name and URL or by visiting the site and using the trackwheel menu to select Add Bookmark to bookmark the current page. Neither of these options is too appealing.

The del.icio.us service provides an excellent way to manage your device bookmarks from your desktop computer, where typing isn't quite the chore it is on your handheld. Just assign another special del.icio.us tag to all the links that you'd like to use on your BlackBerry—"myhandheld"—for instance (see Figure 4-43).

Figure 4-43. Device bookmarks in del.icio.us

After setting up your links with your special tag, use the following URL on your device, where *user* is your del.icio.us username and *myhandheld* is your special tag.

http://del.icio.us/html/user/myhandheld/
?extended=body&tags=no&rssbutton=no

You can just type it in your device bookmarks or, better yet, post it to del.icio.us from your desktop browser and then go to *http://del.icio.us/html/user* and bookmark it from there. Once you've bookmarked the URL, you can go to that link to get a very nice handheld view of your bookmarks without typing them on your device (see Figure 4-44).

Figure 4-44. The same links on a BlackBerry

More del.icio.us

You should also check out Populicio.us (*http://populicio.us/newlinks.html*), which shows popular links over time on del.icio.us. And, if you're a Firefox user, be sure to look at Foxylicious, which syncs your del.icio.us bookmarks with your Firefox bookmarks. See *http://dietrich.ganx4.com/foxylicious/*.

HACK #47 Use Backpack as Your Mobile Workspace

Add notes, to-do lists, and more directly from your device.

Have you ever tried to update a site using a form in your browser like many web applications require? For example, try updating a wiki page from your device. Usually it is not that easy, because most web sites were designed with a computer's browser in mind, not your handheld's.

The Backpack web service (*http://www.backpackit.com*) makes updating web sites easy using email. You just send the content you would like added to a secret email address, and the text is added to your page immediately.

What Exactly Is Backpack?

Backpack is a web service that allows you to get organized by storing your to-do lists, notes, photos, and files online. You can set up reminders that are sent via email. You can optionally share your content with others. The feature that makes Backpack so great for the BlackBerry is the ability to add content to your pages by sending emails.

To use Backpack, you need to create an account. Best of all, there is a basic account that is free and requires no credit card to sign up. Once you choose a username and register, you'll be directed to your newly created section of Backpack, accessible using the URL: *http://username.backpackit.com*.

Click the Make a New Page button on the right to add another page to Backpack, as shown in Figure 4-45. Add a title and some content and click Create. You'll be taken to your new page, where you can use Backpack's very spiffy web interface to add lists, notes, files, images, and links to your page.

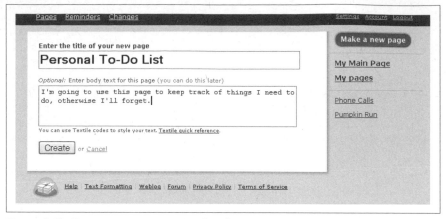

Figure 4-45. Creating a new page in Backpack

Send Email to Your Pages

Each page in Backpack has an email address. Look at the bottom of your newly created page, as shown in Figure 4-46. This particular page has an email address of *lewis00lily@dmabe.backpackit.com*.

Figure 4-46. A page and its email address

You can send any email to this address and it will appear on your page almost instantly. This turns out to be perfect for those moments when you think of a good idea and you need to write it down before you forget. Just send an email to your page, and then access it from your computer later. Figure 4-47 shows a new message from my device, and Figure 4-48 shows how it appears on my page.

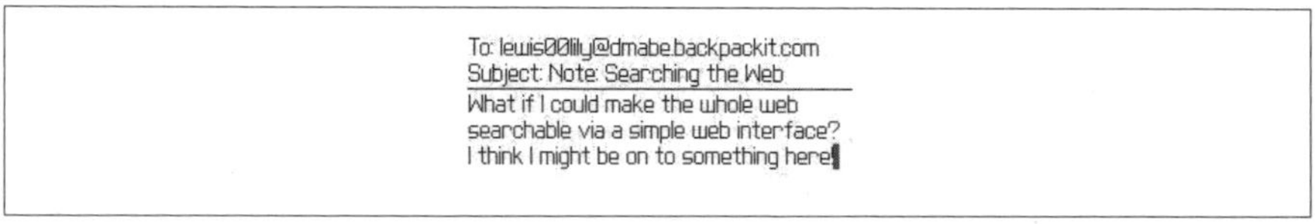

> To: lewis00lily@dmabe.backpackit.com
> Subject: Note: Searching the Web
> What if I could make the whole web
> searchable via a simple web interface?
> I think I might be on to something here

Figure 4-47. Adding a note to Backpack

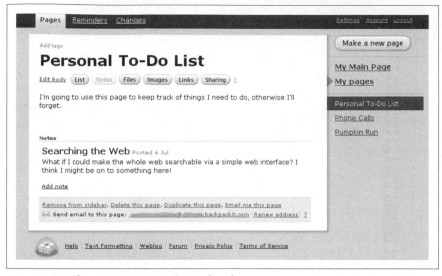

Figure 4-48. The note appears on the Backpack page

Have Reminders Sent to Your Device

Backpack also allows you to set up reminders that can be sent to you by email. The interface for adding reminders in your desktop browser is sleek and usable. It makes clever use of JavaScript using a method called Asynchronous JavaScript and XML (AJAX). Figures 4-49 and 4-50 show setting up a reminder and the email that is sent when it's time. Notice that the formatting of the email allows you to easily use filters [Hack #30] to create a vibration or audible alert when a reminder is received. If you use a calendaring application that supports the iCal standard (for example, Apple's iCal and Mozilla Sunbird), you can subscribe to your reminders so they appear within your calendar program automatically.

Figure 4-49. Adding a reminder in your desktop browser

```
From: Backpack                        ▲
Subject: REMINDER: Pick up the kids
REMINDER: Pick up the kids

You can edit, snooze, or remove this
reminder at:
http://dmabe.backpackit.com/reminders
                                      ▼
```

Figure 4-50. A reminder email as it shows up on your BlackBerry

Access Backpack from Your Device

One of the nice things about Backpack is its ability to access your pages from a mobile device; just append */mob* onto the end of your Backpack URL like so:

> *http://username.backpackit.com/mob*

You can create reminders, new pages, to-do list items, and notes by using this trimmed-down HTML interface in your BlackBerry Browser.

> You can also have the contents of a page emailed to you from the mobile or standard interfaces. Backpack uses a pseudo-wiki style formatting that provides shortcuts for creating new list items and links among other things. You can email the Backpack documentation to yourself and file it in a folder, so you always have access to a good "refresher" course on how to use these shortcuts to produce the markup you'd like in your page. You could also have the Backpack page emailed to you—the email will contain the actual markup Backpack uses to display the page.

HACK #48   Use Gmail as a Spam-Catcher

You started adding forwarding mail to your BlackBerry's email address, and now all those spam messages you were getting are going to your BlackBerry. Keep your BlackBerry from turning into a vibrating advertisement platform for pornographers and discount medications.

Some of us have superfancy spam filters based on Spam Assassin, TMDA, and other content filters. We forward copies of all scrubbed email to our BlackBerry address, or we let the BWC or BIS pull the mail from our mail server. This cuts down dramatically on spam and even viruses, Trojans, and other crud that you don't need to see.

Other users don't have server-side filtering and instead rely on client-side or desktop-based methods to scrub and sanitize email. The most common among these are the *Flag as Spam* filters in Mozilla Thunderbird and other email clients. So while you're filtering your exposure to spam at your local workstation, you're still pulling it down and letting your software decide the spammy-ness of email and acting on it there. This also means that the spam and such stay on the server for delivery to your BlackBerry.

You can roll your own spam filter of sorts using a Google Gmail account. By making Gmail your final destination for emails, and forwarding all your other accounts there, you can set Gmail to trap the spam and send only messages you want onto your BlackBerry address for delivery.

You don't even have to tell anyone you use Gmail, and you can continue using your contaminated primary email address just like you always did, but now you have a huge 2 GB+ archive of all email sent to your BlackBerry (sure beats the 10 to 20 MB T-Mobile USA gives BlackBerry users on the BIS/BWC), allowing for a complete archive of all emails ever received. There are a variety of ways to check your Gmail [Hack #29], and there's nothing that says you have to put your Gmail account as the From: line in the Black-Berry. Instead, configure BIS/BWC to use your primary email address as the From: address, configure your primary email server to forward everything to your Gmail account, and then configure BIS or BWC to retrieve all your mail from Gmail via POP3 (or simply configure Gmail to forward all your email to your BlackBerry's email address, such as *username@tmo. blackberry.net*). Once you've done this, get ready to enjoy the sweet sound of spammers hitting the bit bucket!

Hack the Hack

To make this even more elegant, you can use the filters in Gmail to stream-line operations. I like to keep an archive on Gmail, so I forward the email and cache a copy locally.

1. Log into Gmail and click Create a Filter.
2. Specify "in:inbox" for the Has the Words criteria, and click Next Step. Ignore the error message that pops up.
3. Check the box labeled Skip the Inbox (Archive It).
4. Specify your BlackBerry email address under Forward it To:.
5. Click Create Filter.

If you need to, you can also create a filter that saves you from seeing messages that you yourself send:

1. Log into Gmail and click Create a Filter.
2. Specify your BlackBerry's From: email address for the From criteria, and click Next Step.
3. Check the box labeled Skip the Inbox (Archive It).

4. Specify Apply the Label: Sent From BlackBerry (or whichever label you choose).

5. Click Create Filter.

This will keep messages you've authored on your BlackBerry from being sent to you (you already have a copy on your BlackBerry for your reference).

By making Gmail your filtering, archiving, all-purpose deluxe mailbox, you don't have to sift through spam emails all day on your BlackBerry. You can enjoy the benefits of having Google's technology for indexing and searching all of your email without having to switch to using it full time.

—R. Emory Lundberg

H A C K   Use Your BlackBerry Browser to…
#49   There are a variety of eye-popping wireless-optimized sites that provide excellent information from anywhere. Here is a list of useful ones.

When new users first get a BlackBerry, the email function is usually what attracts the most attention. There is no denying that email is what makes the device great. However, the BlackBerry Browser is an excellent program that allows you to access a ton of useful information on the Web. Having been a relatively recent addition to the BlackBerry operating system [Hack #20], even veteran users tend to pass it by not realizing its utility. Some users may have taken the browser for a quick test drive with their new device, only to have their high expectation unmet as they visited a poorly formatted site made exclusively for desktop browsers.

There are an abundance of sites to get very useful information (even entertainment!) using the BlackBerry Browser, but there is no central list where someone can visit to find out about them. There is a lot of trial and error in finding sites that work well on the BlackBerry. This hack highlights the most useful ones and then lists other sites that also work well and provide access to excellent data that you thought was available only on your desktop computer.

Search Google

Google provides a great XHTML interface (*http://www.google.com/xhtml*) for the BlackBerry Browser. When you access a link in a set of search results, Google actually proxies the request on your behalf and returns the page in a format that's more readable on small screens. In addition to search results, you can access other Google services. Its main search page is accessible as well as the Google Local searches (*http://local.google.com*). Google's local search is quite useful for a mobile user to get driving directions, local restaurant locations, and local maps. Google also provides a nice WML version of its search page (*http://www.google.com/wml*) that searches only sites that are formatted in WML. You can even access a WML version of Froogle, Google's shopping comparison service (*http://wml.froogle.com*).

Log In to Yahoo!

Yahoo! offers an excellent version of its site for handhelds at *http://wap.oa. yahoo.com*. This WML version of the site allows you to log in with your regular Yahoo! ID and use a good deal of their services from WAP browser. You can view your stock portfolios, view sports scores, get weather reports, read the news, and even play several WAP-based games.

Stay in the Game with ESPN

ESPN has a great version of their site available to XHTML browsers at *http:// pocket.espn.go.com/*. The front page is updated with the latest story from the regular version of their page. Each sport's main page reflects the desktop version of its page as well. You can get news, standings, statistics, and results from a variety of sports that look excellent on your BlackBerry device.

Control the Universe with pdaPortal

pdaPortal (*http://pdaportal.com*) is just what the name implies—a customizable portal that you can access from your BlackBerry. You can access a ton of information, including RSS feeds in a nice format, a search engine, and even a random site. Probably the best feature of pdaPortal is its link library of well over 600 sites formatted for handhelds, organized by category. pdaPortal keeps track of hits to the various links it has posted on its site so it can organize its links by popularity.

Go Mobile at BlackBerry.com

BlackBerry's version of a mobile portal (*http://mobile.blackberry.com*) is eye-catching; if you haven't seen it before, check it out. It is formatted with SVG (or *scalable vector graphics*) using the tools from Plasmic. Its look and feel resembles that of a Macromedia Flash application on your computer's browser. It provides links to various sites that are very usable on the Black-Berry. It also provides links to games and ringtones to download.

Geek Out with Slashdot

Every geek's favorite news site has a version of its site available for handheld browsers. Point to *http://slashdot.org/palm* to access it. The problem with accessing the main version of Slashdot on your device is all the comments to each article make the site very large to pull over a GPRS network connection. The handheld version of the site contains no comments (although you can choose to view the top five comments for each article), no images, and no sidebars.

Shop at Amazon.com

You can easily access Amazon.com from your BlackBerry. You can browser and buy items and view the status of orders you've made whether you ordered the item on the desktop version of the site or the handheld version. If you're a heavy Amazon.com user, you may want to try the ShopEdge third-party application for a streamlined interface in a native BlackBerry application [Hack #58].

Control Your Home

If you're into serious geekery, the Misterhouse home automation program (*http://www.misterhouse.net*) has a built-in WML interface that you can use to control your home from anywhere. Control your lighting, turn on your sprinklers, detect motion in your driveway—it's amazing all the cool things you can do with this open source software. Throw your BlackBerry Browser into the mix, and you can do all these things from anywhere you have cell coverage.

Additional Sites

There are countless other useful sites to view from your BlackBerry with more and more popping up every day. Table 4-1 lists some of the noteworthy ones.

Table 4-1. Useful sites accessible via the BlackBerry Browser

Web site	Url
National Hurricane Center	*http://www.nhc.noaa.gov/index.wml*
Geek.com	*http://www.geek.com/portable/index.htm*
The Onion	*http://mobile.theonion.com/*
Moviefone—Movies, Showtimes	*http://palm.moviefone.com/*
MapQuest	*http://wireless.mapquest.com/palm/v3.0/index.html*
CNet News	*http://wap.cnet.com/*
PayPal	*http://www.paypal.com*
Wall Street Journal	*http://wap.wsj.com*
MSNBC.com news	*http://mobile.msn.com/pocketpc/news.asp*
Christian Science Monitor	*http://www.csmonitor.com/pda/*
USA Today	*http://www.usatoday.com/avantgo/index.html*
Stock Charts.com	*http://stockcharts.com/avantgo/*

Table 4-1. Useful sites accessible via the BlackBerry Browser (continued)

Web site	Url
Motley Fool	*http://www.fool.com/partners/avantgo/index.htm*
The Street.com	*http://www.thestreet.com/ag*
BBC	*http://news.bbc.co.uk/text_only.htm*
MSN Mobile	*http://mobile.msn.com/pocketpc/*
Wired	*http://www.wired.com/news_drop/palmpilot/*
MSN Lottery	*http://mobile.msn.com/pocketpc/lottery.asp*
FedEx package tracking	*http://www.fedex.com/p1*
555-1212 phone directory/reverse lookup	*http://www.555-1212.com/palm/*

HACK #50 Find Your Way with a GPS

Use the BlackBerry 7520 to get driving directions and location-based services.

With the BlackBerry's constant network connectivity and portability, it was only a matter of time before RIM introduced a device with a *Global Positioning System* (GPS). The 7520 from Nextel has a GPS chip that can be accessed by developers of third-party applications to create location-based services. TeleNav (*http://www.telenav.com*) is a product by TeleNavigation and is first out of the gate with a great offering for any GPS and BlackBerry enthusiast.

TeleNav can help you navigate to a particular address or airport by speaking the directions as you drive! You can also use it to find certain businesses within driving distance of your current location, wherever that may be. Once you select a local business, you can easily instruct TeleNav to calculate the optimal route to get there and give you directions as you drive. This excellent service costs $29.99 a month, and can be billed right to your Nextel monthly bill.

Get Started with TeleNav

When you start TeleNav, you will be asked for your username and password to log onto the service. Your username will be your phone number on the device. Your BlackBerry will sense that a program is trying to access the GPS chip, and will ask whether you'd like to allow the access. This setting can be controlled by going to Options → Location Based Services and setting the Privacy Setting option to Unrestricted.

Once you've logged into TeleNav (see Figure 4-51), you can use the main screen to select the icon that reads "View a map around an address." You can choose from a variety of locations, including "Here," meaning wherever you happen to be at the moment. Use the Enter key to select Here. TeleNav reads the GPS chip to get the current location and then uses your BlackBerry's TCP connection to communicate with its servers to retrieve the map given your current coordinates. A road map is displayed along with your current coordinates, as shown in Figure 4-52. You can use the U, N, K, and H keys to move around on the map and the number keys to set the zoom level.

Figure 4-51. TeleNav main screen

Figure 4-52. Mapping an address in TeleNav

Navigate to an Address

From the main TeleNav screen, choose the Real-Time GPS Navigation icon to navigate to an address. There are a variety of ways to enter the address to which you'd like to go. You can enter an address on your BlackBerry, or even call in the address to an automated attendant (so you don't have to type anything!). TeleNav verifies the address by doing a live lookup and retrieves the GPS coordinates of the location you entered. You can also use the personalized section of *http://www.telenav.com* to enter addresses on your desktop computer to plan a trip ahead of time.

After TeleNav performs the lookup, verify the destination and choose Get route, and then select Start from here. TeleNav accesses your GPS and communicates with its server to determine the best route to take to get to your destination. Once the route is determined, TeleNav puts you in a navigation screen that shows the current road you're on, the next road you should turn onto, the distance until your next turn, the total trip distance, and the total trip time at your current rate (see Figure 4-53).

As you approach your next turn, TeleNav uses a text-to-speech engine that speaks the directions through the Direct Connect speaker on the device.

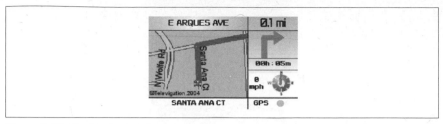

Figure 4-53. Navigation mode

You'll hear it say "Prepare to turn right onto Jones Ferry Road in .3 miles" or "Veer right and merge onto Interstate 40 west bound." Because of the text-to-speech function, you won't even have to look at your device—just keep it parked in the passenger seat. If you do look at your device, you'll see nice and accurate arrow icons showing the type of turn you'll be making next. If the next turn is a 90 degree angle, you'll see an arrow indicating it. If the intersection isn't a right angle, you'll see an icon depicting the direction and angle of the turn—quite useful information for some of the hairy turns you might come across in an unfamiliar location.

If (when) you get off the route that TeleNav has selected, there is an audible double beep, followed by a voice saying "New Route." TeleNav senses you are off course and re-calculates the route. The new route could be a completely new set of driving directions or the first turn could be "Do a U-turn" to get you back on the original route—it all depends on your whereabouts and your proximity to your destination.

Find the Closest Coffee Shop

Have you ever been in a new place and needed a good cup of coffee? I've been in that situation plenty of times and would have paid good money for this feature. TeleNav lets you look up the closest businesses to your current coordinates by type! Go to the Find a Business icon from the main TeleNav screen and select a search point. The most common search point is "Here," but you can also choose a search point from any waypoint you've created or any recent destination. You can also simply enter an address for your search point. You can look up coffee shops, ATMs, restaurants by cuisine, or the closest bed and breakfast. You can even search for gas stations by price per gallon! There is a lengthy list of business types to choose from, and the results are sorted by distance from your search point (probably your current location). You can easily navigate to any of the results with just a couple clicks, as shown in Figure 4-54.

Figure 4-54. Mmmm…coffee

The TeleNav program is very hard on the battery; simply leaving the main screen up on your BlackBerry will access the GPS chip and drain your battery very quickly. Be sure and exit the TeleNav program when you are not using it.

Imagine the Places You'll Go

It only takes a little bit of imagination to invent all sorts of cool tricks to do with a GPS-enabled BlackBerry. For example, imagine a custom application that accesses the GPS coordinates at time periods configurable by the user. It could communicate with a web service over HTTP **[Hack #94]** and upload a custom trip name, username, timestamp, and latitude and longitude coordinates. Using the Google Maps API (*http://www.google.com/apis/maps/*), you could code a web site to map the course of your various trips. The user could set the upload interval to a very small interval for short trips or a much longer interval for extended trips. The program could release control of the GPS between intervals to conserve the battery. The user could send a URL for the trip for others to view. ("Hey, Mom, click here to view the progress of our family vacation in real time. Here's the username and password.")

With all the press that maps and location-based services have been getting recently, having a GPS on your BlackBerry opens up some impressive possibilities for developers.

Free Programs
Hacks 51–59

Where would computing be without free software? Whether you define free as open source, where the source code is available for all to see and modify, or simply freeware, both are important and have their place. If you tried out every commercial program that tickled your fancy, you'd be nickeled and dimed to death. Luckily there are custom applications that come free of charge—if you know where to look. Because RIM chose J2ME as their platform for the BlackBerry, the device is seen as a viable operating system by the millions of Java developers worldwide.

This chapter includes a small subset of the free applications available for the BlackBerry. You can view the night sky [Hack #55], go shopping [Hack #58], and even put your device through the paces [Haok #64] to see how it stacks up. As the BlackBerry third-party application explosion continues, the number of free programs available for the device will continue to grow.

HACK #51 Get Fast and Free Mobile Search

This program provides quick and convenient on-the-go searches of the yellow pages, the white pages, the Web, movies, and more.

Calls to 411 on your BlackBerry can add up to big charges on your monthly phone bill, and even after you've gotten the information you want, there's no easy way to save it for later use. On the other hand, using PC web sites like Google Local from the BlackBerry is slow and inconvenient.

You can also access Google Local using SMS, or the text messaging feature on your mobile phone. Just send your search terms to 46645 (GOOGL on most phones) and you'll get a reply with results from Google. Although this is convenient, most plans charge for each SMS message you sent, so these queries can quickly add up.

Berry 411 provides the best of both worlds: free and convenient mobile search optimized for the BlackBerry. In addition to white and yellow pages with maps and driving directions, Berry 411 provides searches for Google results, movies, and price comparisons. You can dial phone numbers directly from the result screen and add them to your Address Book.

The over-the-air install [Hack #97] is available at *http://www.thebogles.com/ berry411.jad*, and the install for Application Loader is available at *http:// www.thebogles.com/berry411.zip*. The official home page for Berry 411 is *http://www.thebogles.com/Berry411.htm*.

Berry 411 is charityware: users are encouraged to make contributions to a number of charities; consequently, there are no per-search charges with Berry 411.

Set Up Your Addresses

Berry 411 launches from an icon on the Home screen. The first time you run Berry 411, or whenever you select "Edit Addresses," Berry 411 will prompt you for your home and work address, as shown in Figure 5-1. You can enter a Zip Code instead of a city name if you wish.

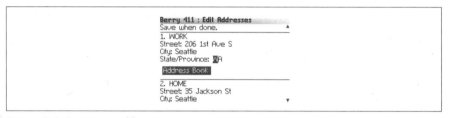

Figure 5-1. Setting up addresses

Search

The main Berry 411 search screen (shown in Figure 5-2) is the hub for all the different kinds of searches. It allows you to cycle between your home, work, and other addresses, to enter your search keywords, and to select the kind of search you'd like to perform.

For example, if you wanted to find the Mediterranean Mix restaurant near your work, you could type in "med mix" and click the trackwheel twice to select Yellow Pages search.

Berry 411 remembers the last location you searched for, so often you don't need to change it. You can toggle your current location by scrolling to the location drop-down menu and hitting the spacebar.

Figure 5-2. Berry 411 Search screen

Yellow and White Pages

The yellow page results page displays the names, phone numbers, and addresses of all the matches. From the results page (Figure 5-3), you can dial the phone number or even add it to your Address Book.

Unlike traditional 411, you can view maps of each result (Figure 5-4) and get driving directions from your currently selected location (Figure 5-5).

Berry 411 displays results from Google Local carefully optimized for the BlackBerry screen. The white pages search does the same thing for people in Google phonebook.

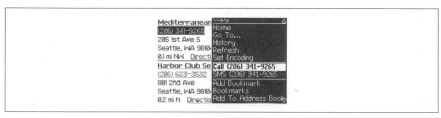

Figure 5-3. Yellow page results

Figure 5-4. Maps

Web Searches

Sometimes you're on the road and need to check out some facts from the Web. Berry 411 makes it convenient to search the Web using Google (Figure 5-6) and to obtain mobile-friendly reference information using MobileAnswers (*http://mobile.answers.com/*).

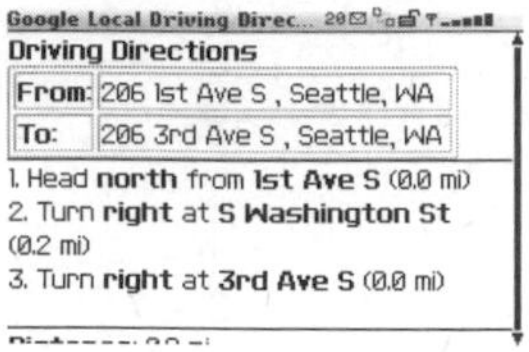

Figure 5-5. Driving directions

Figure 5-6. Google searches

Movies

Suppose you've just finished dinner at a restaurant and want to know where a particular movie is playing. Just type a few words from the movie's title into Berry 411 and select Movies from the menu. If you want to know all the movies playing in your neighborhood, just leave the search box blank. (See Figure 5-7).

Figure 5-7. Movies results

Shopping

When you're in a store, it can be difficult to know whether you're getting the best deal. The Shopping search integrates with Froogle to show you the best prices from the Web for the item you're about to purchase (see Figure 5-8).

Figure 5-8. Froogle price comparisons

Yellow Page Search Tricks

These tricks will help you get more out of Berry 411 yellow pages:

- Override the default search location by including it in your query (e.g., "Antiques Hanover NH" or "school 02139").

- Get driving directions to a specific address by searching for that address in the yellow pages (e.g., "206 1st Ave S, Seattle WA").

Web Search Tricks

You can take advantage of many hidden features when you search Google using Berry 411.

- Search for "weather" using Google to see local weather.

- Use Google as a calculator by entering an expression (e.g., "72 * 15%").

- You can convert back and forth between English and metric measurements. Just enter a size and unit (e.g., "20 miles").

- Search for "wk *word*" to find the wikipedia.com entry for *word*.

- Search for "answers *word*" to find the *mobile.answers.com* entry for *word*.

Extend Berry 411

The author is currently seeking suggestions for new Berry 411 features, and future versions of Berry 411 will support the ability to add your own custom search types by entering a name and URL. Please contact the author with your suggestions, as described on the Berry 411 web site.

—*Phil Bogle*

Store Your Passwords Securely

#52 Every online service including shopping, banking, and news requires an account and password to be established for access. Learn how to avoid having to remember all the different combinations of IDs and passwords.

One of the newest features added to the latest handheld operating system [Hack #20] is Password Keeper. You can see its icon, a vault, in Figure 5-9.

Figure 5-9. New Password Keeper application

Set Up the Password Keeper Password

Password Keeper provides a secure means to store various information such as IDs, passwords, and PINs using AES encryption technology. You can also use it to generate random passwords or to copy a password to the clipboard, which you can later paste when accessing an application through the BlackBerry Browser. When you open Password Keeper for the first time, you are prompted for a password and must reenter it to confirm. The password screen displayed is shown in Figure 5-10. Once you have typed the password, click OK to continue. Password Keeper is ready to use!

> Even if your handheld uses password protection to unlock it, you will be prompted for another password for the Password Keeper application.

Figure 5-10. Setting the password for Password Keeper

Add an Entry to Password Keeper

Now you are ready to add an entry to your password vault:

1. Click the trackwheel and select New from the menu. The Password Keeper displays a New Password screen.

2. In the Title field, enter a brief description that will help you identify what the entry relates to.

3. Enter the username in the next field.

4. Continue to scroll down, and then enter the password associated with that username.

5. Enter the web site used for access under the Website field, if it applies.

6. Under the Notes field, enter any additional information you would like to note about this entry. You can see a completed entry in Figure 5-11.

7. Click the trackwheel and select Save.

8. Repeat the steps above for each account you would like to enter.

 Don't forget! Each time you open Password Keeper you will be prompted for your Password Keeper password!

Figure 5-11. Sample Password Keeper entry

When you want to access a password, just open Password Keeper and scroll down to the desired entry. You can click and select to edit or view the entry. If you are using the BlackBerry Browser to access a mobile online service or application, you can use Password Keeper to copy the vaulted password to the clipboard to later paste into the application. This is done by simply right-clicking the entry and selecting Copy to Clipboard. Once copied, open the online application and paste the password into the proper field.

You can also use Password Keeper to generate random passwords as well as define the criteria for those passwords. Various options are available; open Password Keeper, click the trackwheel, and select Options. The Password Keeper Options are listed.

If you use Desktop Manager to back up or restore your handheld data, your passwords will also be safely backed up along with your other data. Since the data is encrypted, your passwords are safe!

—*Shari Kornberg*

HACK #53 Skin Your Device

Customize the appearance of your 7100 or 72xx series device using themes.

The older BlackBerry devices couldn't dream of being skinned—there wasn't much to skin anyway! Those days of monochrome screens with very low resolution are gone. The new BlackBerry devices have great screens with a ton of colors and much higher resolution. These improvements allow customization for the display in ways that weren't even thinkable back in the days of the 850 data-only devices.

The 7100 and 72xx series devices support *themes*, which are similar to "skins" for other applications such as Windows Media Player and Winamp. By default, the alternate themes are not loaded on your device. You'll have to download them and install them with Application Loader.

Installing the Themes

At the time of this writing, there are four additional themes for your 7100 device in addition to the theme that it ships with. You can download the themes from the following URL: *http://www.blackberrycool.com/bbthemes.zip*.

Extract the zipped archive to a local folder and start Application Loader. Click on the Add button and browse to the directory where you extracted the theme files. Select the *Themes.alx* file and click Open, as shown in Figure 5-12. Click Next to install the additional themes to your device.

Changing Themes

Of course, now is the fun part. Go into your device's settings and choose the Theme option. You should see some new themes available in addition to the default theme that came with your device. Figure 5-13 shows the list of themes as they are displayed on a 7100g device from Cingular. The default theme has the "(List)" string on the end of its entry.

To change to a particular theme, just select it and use the trackwheel to choose Activate from the menu. Figure 5-14 shows the O2 theme that looks quite nice, although there are some reports that this theme slows some devices down noticeably.

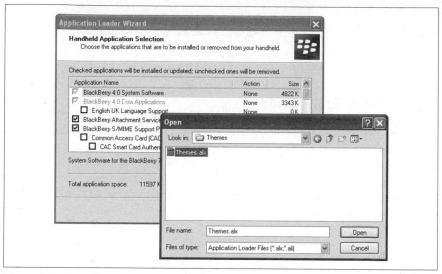

Figure 5-12. Installing the themes

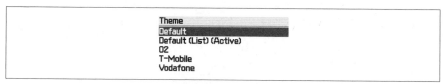

Figure 5-13. Additional themes on a 7100g

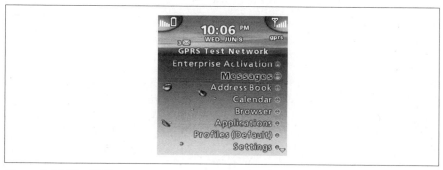

Figure 5-14. The 02 theme

Despite the ordinary name, many users who have upgraded from previous BlackBerry models will find the theme called Default most comforting, as it makes your 7100 appear almost identical to the older devices (see Figure 5-15).

Figure 5-15. The "Default" theme

 This "themeability" was requested by carriers and really isn't fully supported by RIM. There is no documentation on how to create your own theme, and you can't download any themes from RIM's web site. The carriers wanted to be able to customize their versions of the devices.

HACK #54 Run a Stress Test

Use this free program to put your device through a stress test.

Research In Motion has released a variety of devices for a variety of networks. The resources available to each handheld model vary and, of course, over time the hardware is getting faster with more capacity. You can compare your device's performance against the average performance of other models by using a free program called BenchmarkMagic.

Install BenchmarkMagic

You can download BenchmarkMagic from the author's web site at *http://www.software-for-blackberry.com*. You'll be redirected to the popular handheld software site Handango, where you can download an archive suitable for installing using Desktop Manager and Application Loader. There is no over-the-air install available at the time of this writing.

Once you install BenchmarkMagic, run it by clicking on the icon that resembles a processor on the Home screen (see Figure 5-16).

You'll be presented with the main BenchmarkMagic screen showing a variety of BlackBerry models and the average performance scores they achieve using the tests this software performs.

Figure 5-16. The BenchmarkMagic icon on the Home screen

Put It Through the Wringer!

When you're ready to run the test, use the trackwheel to access the menu and choose Run Test—then watch the sparks fly! The program creates several polygons using the BlackBerry's graphics library and prints them to the screen (see Figure 5-17). This hardware-intensive operation repeats several times, making a good test of your hardware.

Figure 5-17. The benchmark test in action

Once complete, your handheld's score will be displayed on the screen, showing the number of polygons your device was able to produce in a second and the number of CPU operations it was able to perform in a second. Click OK to return to the main benchmark screen, as shown in Figure 5-18 where you'll see the score from the test you just ran compared with other BlackBerry devices.

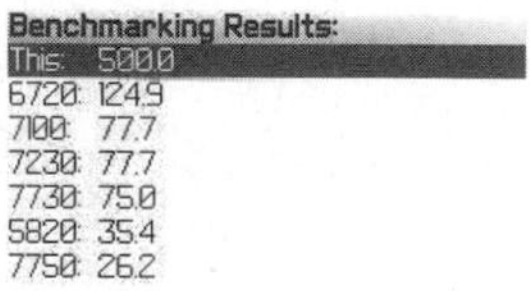

```
Benchmarking Results:
This    500.0
6720:   124.9
7100:   77.7
7230:   77.7
7730:   75.0
5820:   35.4
7750:   26.2
```

Figure 5-18. The results of your test

You'll see that my device scored off the charts! It's several times faster than the next most powerful device. Is this the latest prototype BlackBerry with the latest, fastest processor available? Not exactly. Actually, I cheated—I ran this test using the BlackBerry Simulator **[Hack #93]** that runs on a Windows machine. These handhelds are no match for my Intel Pentium 4 processor!

Install a Planetarium

HACK #55

This free application provides everything an amateur astronomer needs. It even transforms your device into a powerful telescope. Well, not really.

Do you ever look up at the night sky in dazzling amazement? Do you try to impress your friends by pretending to identify some obscure constellation even though you have no idea what you're talking about? Well, this free application can make anyone an expert.

MicroSky is a freeware program that is made for J2ME phones and includes a special version designed to work on the BlackBerry. You can view images of the night sky with the constellations already labeled. You can view the night sky from your location's perspective. MicroSky allows you to view what your night sky will look like at a certain time, so if there is a certain constellation you'd like to view, you can know when it is available for viewing.

MicroSky uses a *SkyServer*, which provides data to the public from the Sloan Digital Sky Survey project. The data from the project provides access to almost 200 million objects in our sky. You can even download a portion of the main database and create your own SkyServer.

Download the Program

MicroSky offers an over-the-air installation (which, of course, is the easiest way to install), as well as *.alx* and *.cod* files suitable for installing with Application Loader using your USB connection. Access either installation here: *http://www.upto.org/microsky/blackberry/*. If you're installing MicroSky over the air, just click on the *MicroSky.jad* link on the top of the page. If you need to install using Application Loader, download both the *MicroSky.cod* and *MicroSky.alx* files and put them in a directory on your computer. Bring up Application Loader and browse for the *.alx* file you just downloaded.

Start Your Career as an Astronomer

You'll first need to create an account on the SkyServer to use the application at *http://www.upto.org/microsky/*. Although it is a little easier on the thumbs to create your account using a desktop browser, the registration page works just fine in the BlackBerry Browser. You'll need to give your first and last name and choose a username and password to log on with. There are three mobile locations at the end of the registration form that you are required to fill in. Select the location that is closest to you for the first field, and then choose other locations that you might be interested in for the remaining two. You'll be able to toggle between these locations on your device to view the sky from the perspective of each location.

After you've set up your account, access the MicroSky icon from your Home screen (it looks like a hurricane). Use the trackwheel to access the Logon option from the menu. Enter your newly created username and password. Use the trackwheel to access the menu and choose the >Go Now option, as shown in Figure 5-19.

Figure 5-19. Choosing an option from the menu to view the sky

You are presented with an amazing image of the sky from the perspective of your first location you chose when you registered your account. Figure 5-20 shows an image I captured using a BlackBerry 7290. You can also change your location by entering GPS coordinates.

Figure 5-20. The night sky from North Carolina

The crosshairs in the middle of the image can be moved up and down and side to side. You can recenter the image according to crosshairs' location by choosing the >Center Position option from the trackwheel menu.

Browse by Constellation

MicroSky comes with a list of constellations that you can choose from to view the sky with a particular constellation in the center of the image. Choose the >Go Constellation option from the trackwheel menu to view the list. The list will appear to be truncated at first—you'll need to use the More Entries option on the menu to view the next set of 20 constellations when you get to the end of the list. Highlight the constellation you'd like to view, and choose Select from the trackwheel menu to view it (see Figure 5-21).

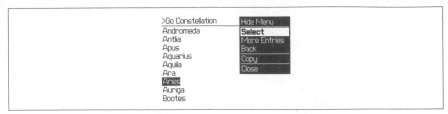

Figure 5-21. Choose a constellation to view

You can zoom in, out, or reset the zoom using the >Settings option and selecting zoom. To view the details of what you're viewing, choose Details from the trackwheel menu as shown in Figure 5-22.

Figure 5-22. Details of your current view

Get Your Bearings

To determine where you're currently looking in the sky, MicroSky provides a graphical display of the *altitude* and *azimuth* of the current image given your current location. These are represented by the half disks on the bottom and side of the image as shown in Figure 5-23.

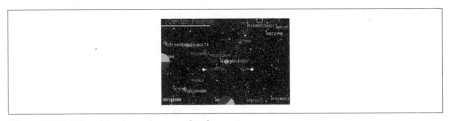

Figure 5-23. The altitude and azimuth of your image

The altitude is represented by the disk on the bottom of the image and the azimuth is represented by the disk on one of the sides. The background will appear gray and the value will appear in yellow. When the image you're currently viewing is below the horizon, the altitude disk will appear completely gray. You can use the Details option as displayed in Figure 5-22 to get the exact values for the altitude and azimuth.

For more hacks that will help you better observe the night sky, be sure to see *Astronomy Hacks* (O'Reilly, 2005).

HACK #56 Rein In Your Backlight

The default backlight settings are good for your battery, but not good for reading in the dark.

You're in a car at night and your BlackBerry device just gave you an alert that you received an important message that requires your immediate attention. It turns out to be a fairly lengthy email and as you read it, your device's backlight hits its built-in timeout after just a few seconds of inactivity. What a pain! You'll soon start going through meaningless trackwheel gyrations just to reset the backlight timeout.

Unfortunately, there is nothing in the default BlackBerry operating system to control anything about the backlight. Never fear, however, because there is a simple-to-use, free piece of software that allows you to control various backlight settings. It will even let you keep it on all the time!

Install and Use BBLight

The BBLight software is one piece of the open source BlackBerry Tools project (*http://sourceforge.net/projects/blackberrytools/*). Unfortunately, there is no over-the-air install for this application, so you will have to download the zip file for installing with Application Loader. When you install BBLight, it places a light bulb icon on your Home screen, as shown in Figure 5-24.

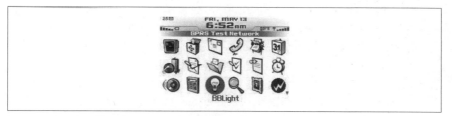

Figure 5-24. The BBLight icon on the Home screen

Backlight Settings

Go to the BBLight program from the Home screen. Enable the checkbox next to Activate at Startup, as shown in Figure 5-25. From this point forward, when you activate your backlight, it will stay on until you turn it off—the perfect setting for reading in the dark. Of course, it is easy to forget and leave the backlight on. Beware—doing this will eat your battery alive!

There are other settings to control when your backlight is turned on. For non-7100 devices, you can have the backlight turned on whenever you click any button on your BlackBerry so it acts much like a typical mobile phone (the 7100 series devices do this already). There are also settings for turning it on every time you remove your device from its holster or connect the external power source, or by time of day.

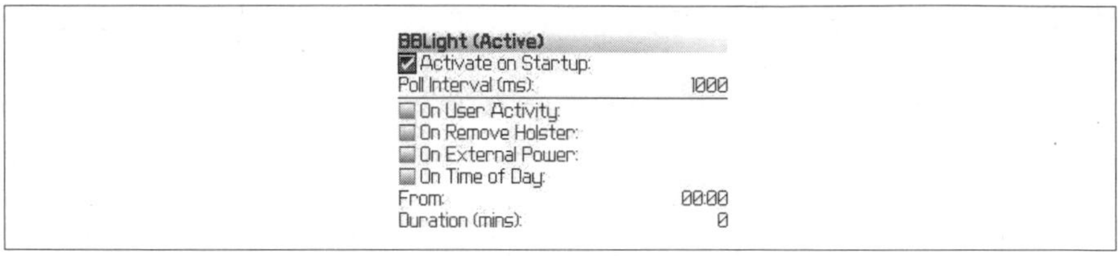

```
BBLight (Active)
☑ Activate on Startup:
Poll Interval (ms):              1000
▣ On User Activity:
▣ On Remove Holster:
▣ On External Power:
▣ On Time of Day:
From:                           00:00
Duration (mins):                    0
```

Figure 5-25. BBLight settings page

Use Multiple Reply-to Addresses

HACK #57

By default, the BlackBerry doesn't allow you to use more than one reply-to address. This application allows you to create a list of 10 to choose from with each message you compose.

When you set up the BlackBerry Web Client, you specify an email address that will show up in the From field of your outgoing messages. This address can be anything you choose. The only limitation is that the From field you specify is used for all messages you send through the BlackBerry Web Client and can only be changed by using the web interface, which isn't accessible from your BlackBerry.

Fortunately, there is a free program that allows you to change the *reply-to* address for each message you send through the BWC. Note that this will not change your *From* address, but you can specify that replies to messages you send go to any of a maximum of 10 addresses you configure on your device.

Install BBReply

BBReply is part of the BlackBerry Tools project hosted at Sourceforge (*http://sourceforge.net/projects/blackberrytools/*). There is no over-the-air install for BBReply, so you will have to download the zip archive and install it by using Application Loader.

Once installed, BBReply puts an icon on your Home screen, as shown in Figure 5-26.

Figure 5-26. The BBReply icon on the Home screen

Use BBReply

Clicking on the icon will bring up the BBReply program. Use the trackwheel to bring up the menu and choose Add to add a new reply-to address. Figure 5-27 shows the screen you'll use to configure a new address.

Figure 5-27. Adding a new reply-to address in BBReply

Once you've added an address to your list, when you compose a new message, you'll get an additional option on the menu that appears after you click the trackwheel, as shown in Figure 5-28.

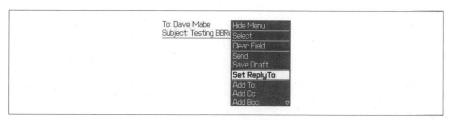

Figure 5-28. The new option on the message menu

Choosing the Set ReplyTo option from the menu will present you with the list of addresses you've defined in BBReply. You can also add reply-to addresses from this screen right in the middle of writing your message.

After you've selected your reply-to address, it will appear between the To and Subject fields in your new message. You can then send the message as you normally would. Figure 5-29 shows a message sent with a custom reply-to address using this method.

Note that the name that you give the address in BBReply not only serves as a way to identify the address when you are composing a new message, but it also appears in the friendly name portion of the email address that gets sent. Be aware of this as you add addresses in BBReply.

Figure 5-29. Headers for a message sent with a reply-to address using BBReply

The reply-to address only specifies the addresses to which replies should be sent, and most email programs respect this header when the recipient presses the Reply button. The Reply-To is different from the From address, which you won't be able to change unless you go into the BWC web interface. So even if you set a Reply-To address, the recipients will always be able to see whatever From address you set up in the BWC web interface.

HACK #58 Shop Amazon.com

This program allows you to search, add to cart, and purchase from your device.

Although you can access the Amazon.com site through your BlackBerry Brower, the interface is a little clunky. The WAP version has no images, so you can't see something before you purchase it. Also, as with a lot of WAP sites, it just doesn't feel like an easy-to-use application. The HTML version of the site is not very usable on a handheld device.

There is an excellent product called ShopEdge from a MAQ Software that makes shopping at Amazon.com a breeze. There are versions for BES and WAP users and, best of all, it's free.

There are convenient options for downloading ShopEdge over the air or through your USB cable using Desktop Manager and Application Loader. The over-the-air install is available at *http://shop.edgeq.com*, and the install for apploader is available at the following URL: *http://www.edgeq.com/ download.aspx?application=ShopEdge&platform=blackberry.*

Search Amazon.com

The first time you run ShopEdge, it will ask you which version of the store you'd like to shop at (U.S., U.K., etc.), and then if you installed the WAP version, it will ask for your carrier information. Once you've configured it, searching Amazon.com is simple—much easier than shopping via the HTML or WAP interface. The Search screen (shown in Figure 5-30) is clean and simple. Just like shopping at Amazon.com from your PC, you can search all products or refine your search to a specific category.

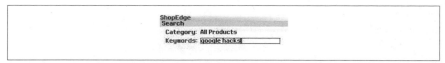

Figure 5-30. The ShopEdge Search screen

Once you've entered your keywords, click the trackwheel once to bring up the menu and choose Search. As the program searches for matches, a nice screen appears with a progress bar before the results (see Figure 5-31) are returned.

Figure 5-31. The Results page

Once you find an item in your results that you'd like more information on, click the trackwheel and choose Get Details from the menu. The Amazon.com image for the product appears on the Details page (Figure 5-32), along with the average customer rating, price, authors, and the availability status.

Figure 5-32. The Details page for an item at Amazon.com

Purchase an Item

There are a couple different ways to purchase an item you've found using ShopEdge. You can click the trackwheel and choose Buy Now on the menu. This spawns your default browser and sends you right to a stripped down Amazon.com page for the item, allowing you to purchase the item directly from your BlackBerry.

You can also add the item to your cart, which is stored within ShopEdge on your device. When you're ready to check out, you can choose to purchase the items in your cart directly from Amazon.com on your device by choosing the Transfer to Amazon option while viewing your cart.

If you prefer to purchase the items from your computer, there is a nifty option to email your cart to your email address. From the My Cart screen, choose the E-Mail Cart option from the menu, as shown in Figure 5-33.

Figure 5-33. The My Cart screen and options

The E-Mail Cart option sends an email to the address you've specified in the ShopEdge settings with a URL on Amazon.com. The email message will look similar to the following.

> Dear Customer,
>
> Please click on the URL provided below to complete your purchase:
>
> https://www.amazon.com/gp/cart/aws-merge.html?cart-id=103-6369303-3891057%26associate-id=maqsoftware-20%26hmac=ku30v5roYPNciqbyR4twNSAf3Wc=%26SubscriptionId=1B4 HD1RG6FYBS7F59BR2%26MergeCart=True
>
> Your cart contains the following items:
>
> 1) Google Hacks, 2nd Edition (Hacks)
>
> In case of any enquires, please send an e-mail to products@maqsoftware.com
>
> Thank you.
>
> ShopEdge team - MAQ Software

When you click on the link, it will take you to Amazon.com and merge your ShopEdge cart with your Amazon.com cart. From there, you can purchase the items just as you would from your computer.

The observant Amazon.com geek will glance at the URL above and realize MAQ Software's revenue stream. Their Amazon.com Associate ID is embedded in the URL, meaning they'll get a kickback for every item ordered through their product. See *Amazon Hacks* for more details on the Amazon Associates program.

HACK #59 Manage Your eBay Auctions

This nifty program allows you to search, bid, and monitor items on eBay.

eBay provides a nice WAP interface for you to access many features of the service from your BlackBerry Browser. It allows you to search for items, make bids, and access auctions you are watching, but it is limited in a few ways. The interface is a little clunky (it is a WAP interface, after all) and you can access only certain eBay features—for example, you can't save or view previous searches you have made.

There is an excellent program from Abidia called Abidia Wireless (available for a one-time charge or a monthly subscription) that gives you more access to eBay features in a nice interface for your BlackBerry. There are versions for BES and non-BES BlackBerry users.

There are a number of ways to install Abidia Wireless. You can install it over the air from Abidia's WAP interface at *http://wap.abidia.com*. There is also a download for Application Loader if you want to install over your USB cable at *http://www.abidia.com/install/pc*. Another unique way that Abidia allows you to install their software is using SMS. Go to *http://www.abidia.com/install/sms* and enter your wireless phone number and carrier, and Abidia will send you an SMS message with the download URL to the over-the-air install. Just access the URL directly from the text of the SMS message, and you're on your way.

Configure Abidia Wireless

You can access eBay through Abidia Wireless anonymously, but you'll get a better experience if you enter your eBay username and password. Click the trackwheel to bring up the menu and select Settings. Enter your eBay username and password, as shown in Figure 5-34.

If you are using the WAP version of Abidia Wireless (as opposed to the BES version), you'll need to configure your APN settings specific to your carrier. Abidia provides the settings for most every carrier in the quick start guide, available at *http://www.abidia.com/srvc/apps/doc/wire.rim.3.0.0.pdf*.

Figure 5-34. Entering your eBay username and password

Use the Program

One of the main differences between Abidia Wireless and the WAP interface provided by eBay is the ability to create and save searches. To search eBay, click on the trackwheel in the main Abidia screen (see Figure 5-35) and select New.

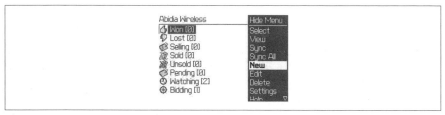

Figure 5-35. Creating a new search

As shown in Figure 5-36, enter your search terms in the Keyword(s) field and optionally choose among the three choices for refining your search, which should be self-explanatory if you've ever bought anything on eBay before.

```
Search Entry | Abidia Wireless
Keyword(S): WRT54G
Search Options:
☐ Completed Items Only
☐ Buy It Now Only
☐ Title and Desc
```

Figure 5-36. The Search settings

Use the trackwheel to access the menu and select Done to save your search. You'll now see an additional item on the main Abidia screen that corresponds to the search you just entered. To actually execute the search, highlight the search from the main screen and use the trackwheel to access the menu and choose Sync. A progress screen appears, showing the status of the data connection as it is being made. When the search is complete, you'll return to the main screen and your search will show the number of items that matched your search terms. Use the trackwheel to choose Select from the menu to view all the search results. You can view an item (see Figure 5-37) and make bids on an item by selecting it and using the trackwheel menu.

Figure 5-37. Viewing an item

Interpret the Search Results

Different items on eBay are displayed in different ways in the search results in Abidia. Regular items will appear with the current bid price at the beginning of the line. Items in your search results beginning with C are completed auctions. Items that begin with the string "bin" are Buy It Now only items, meaning there is no auction for the item, just a set price that the seller is offering it for. Items with two prices right at the beginning are items that are up for auction and have a Buy It Now price. Figure 5-38 shows my watch list with each of the types of items in it.

Figure 5-38. Different item types

Sync with My eBay

Probably the best feature of Abidia Wireless is its ability to stay in sync with the My eBay (*http://my.ebay.com*) portal that you usually would access from a computer. From the main Abidia screen, choose the Sync All option from the trackwheel menu to synchronize between all the sections of My eBay. This gives you access to items you're bidding for, items you're selling and have sold (or unsold), items in your watch list, and items you've won and lost.

This program is excellent for using eBay on the go. When eBay sends you an email saying you've been outbid on an item, you can go right into the Bidding section of Abidia and bid again from your device. You can also add an item to your watch list from your device, and then easily view the item in more detail when you return to a desktop computer.

Shareware Apps
Hacks 60–71

There's nothing like free enterprise to spur innovation and fill the needs of paying customers. In places where the BlackBerry comes up a little short (there are a few—not many, but a few), there are a growing number of third-party developers ready to fill the void. This chapter showcases the entrepreneurial spirit of the BlackBerry application developer. Most of these providers are small businesses (some even a single person!) that have found a nice niche with the BlackBerry. RIM's loyal customer base makes a nice, motivated target audience for third-party applications. If you've got a need, there's a good chance there is a BlackBerry application that can fill it. If there's not a program that does what you're looking for, at the current rate of development, just wait a couple weeks! This chapter highlights some useful third-party programs to get you started: a real-time stock quote program [Hack #70], a program to modify Office documents [Hack #68], an alternate web browser [Hack #66], and an alternate email program [Hack #67].

HACK #60   Display a BlackBerry Today View

Use this application to display your recent emails, upcoming appointments, and other pertinent information on a single screen.

Just as Microsoft Outlook can display an overview of current items such as unread messages, today's appointments, and tasks, you can use an add-on application to do the same on your BlackBerry. The software is called Pocket-Day and is made by a company called Cross River Systems. The product has continued to mature, and additional features have been added to make this an outstanding and very useful product well worth the modest price.

Install PocketDay

Like most applications lately, PocketDay offers convenient over-the-air install. PocketDay is available at *http://www.crossriversystems.com/PocketDayBB.htm*. Because PocketDay provides an intelligent set of default options, you'll be up and running in no time. Just execute the PocketDay application from your Home screen and enter your license to register the product. You will then be presented with a polished summary of your current items from the various applications on your BlackBerry, as shown in Figure 6-1.

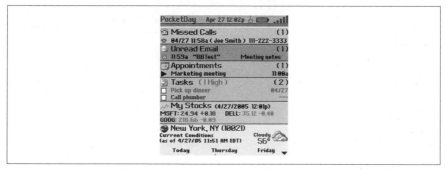

Figure 6-1. The PocketDay screen

PocketDay Features

PocketDay aims to be your main application on the BlackBerry, where you spend most of your time. Right away you'll notice that PocketDay gives the time and date, your battery status, and wireless coverage indicator—all the things you've come to expect from your BlackBerry Home screen. But on the same screen you can see missed calls, unread emails, upcoming appointments, tasks, a weather report, and stock quotes, as well as the underlying details of many of these items. Scroll to any item displayed and use the Enter key to access the particular item right from within PocketDay. You can even delete items such as email messages from within PocketDay.

If you don't like the way PocketDay looks, you can customize its appearance using a number of options. You can control the maximum number of items that are displayed within each category and whether some items appear at all. You can also reorder any of the categories. There is also an option that tells PocketDay to start up when you turn on your device. Enabling this setting allows you to use PocketDay as a virtual replacement for your BlackBerry's native Home screen.

Use Advanced Features

Too many third-party applications for the BlackBerry are simply minor rewrites from a version of the software designed for a standard mobile phone. The best part about PocketDay is that it looks and feels like what a BlackBerry application should feel like. There are hotkeys to each built-in application (which you can modify if want to). You can customize the colors and fonts for each category. You can even assign hotkeys from within PocketDay that take you to other third-party applications.

Figure 6-2 shows how you can add a shortcut key that is accessible from within PocketDay. As you can see from the pick list, you can choose any application you've installed on your BlackBerry. In this example, I assigned the shortcut key "u" to the BlackBerry's Calculator program (the "c" shortcut is already taken; it goes to the Compose Mail screen). After saving these settings, you can simply hit the "u" key to be taken directly to the Calculator program, as shown in Figure 6-3.

Figure 6-2. Adding a program shortcut into PocketDay

Figure 6-3. Using a PocketDay shortcut to go to the Calculator application

> Before assigning your own shortcut keys in PocketDay, be sure to check which shortcut keys are available manually. PocketDay doesn't check whether the shortcut key has already been assigned to another function, so you can end up with more than one function mapping a hotkey.

Use SpeedDial in PocketDay

The speed dial keys that you assign in the built-in Phone application [Hack #18] aren't available within PocketDay, so it provides its own speed dial functionality. Use the trackwheel to bring up the menu from within PocketDay and choose SpeedDial. Figure 6-4 shows the SpeedDial screen that appears. To assign a SpeedDial entry, select your desired shortcut key, click the trackwheel, and select Set SpeedDial. You'll be presented with a list of your contacts where you can choose one and then choose the specific number you'd like to assign.

If a specific contact to which you wish to assign a speed dial entry is not showing up in the list, be sure the contact has an email address. Due to a quirk in the BlackBerry operating system, only contacts that have email addresses will appear in your list.

Figure 6-4. PocketDay SpeedDial Settings

HACK #61

Sync Your BlackBerry to Your Mac

Mac users don't have to watch BlackBerry users from afar; they can also join thumbs with the rest of the world by using software to keep their Mac and BlackBerry in sync.

The BlackBerry's software is clearly focused on the Microsoft Windows desktop components and infrastructure: the only software that RIM supplies for synchronization, firmware updates, and application loading requires Windows. Mac users aren't left entirely in the cold, however, thanks to some ingenious software from PocketMac (*http://www.pocketmac. net/*), called PocketMac for BlackBerry. While you will not (at the time of this writing) be able to update the firmware of the BlackBerry or install packages that require the use of the Application Loader, you can happily sync your BlackBerry to a variety of PIM for the Mac, including DayLite, iCal and Address Book, and Microsoft Entourage.

PocketMac for BlackBerry works with a password-protected BlackBerry just fine and prompts you for the password before sync operations (see Figure 6-5).

Figure 6-5. PocketMac for BlackBerry prompting for a password

One of the nicest things about this software is that it allows you to sync against so many applications and gives you fine-grained control over which parts of your BlackBerry sync against which applications: if you really wanted to, you could sync your Contacts against Address Book and your Calendar against Entourage. You can decide if you want your Memopad to get synced against the old school Stickies application, Entourage, or Day-Lite notes. Figure 6-6 shows one of PocketMac's configuration screens.

 If you haven't taken a look at DayLite from Market Circle (*http://www.marketcircle.com/*), you really should. It is easily the most feature-rich of any PIM on the Mac and is in fact a full-fledged Customer Relationship Manager (CRM). You can do just about anything with the darned thing; organizational junkies should at least check out the demo of it. You will probably fall in love with it.

The BlackBerry's Memopad is a great place to store lists and notes for things you need to keep on hand. iSync cannot sync notes between devices, but PocketMac can. So, using a BlackBerry and PocketMac is your best option to keep your notes in sync at all times (see Figure 6-7).

While the idea of syncing isn't at all foreign to Mac users (we've had the ability to sync all sorts of devices to our Macs for quite a while), there are some tweaks that you may need when your BlackBerry has been naughty or is otherwise giving you a hard time.

PocketMac has the means to help you rebuild your BlackBerry with the data on your Mac by using the PocketMac Batcave, hidden away in */Library/ PocketMacBB/*.

Open the Advanced Preferences application from that directory, and you will see all kinds of options.

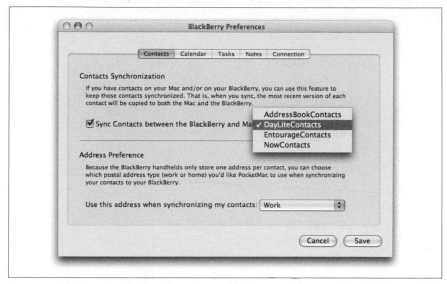

Figure 6-6. Delegating which applications to use for various parts of the BlackBerry PIM

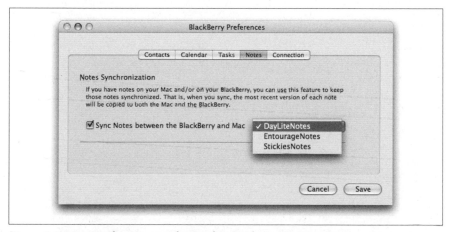

Figure 6-7. Notes and memos can be synchronized

You can have the data in your desktop PIM completely obliterate the contents of your BlackBerry (see Figure 6-8), and it will never sync the data. It will erase the Contacts, calendar appointments, tasks, and notes, and then copy the items from your Mac right onto the BlackBerry.

Another useful option for iCal users is to only sync things against one calendar—this way, you can keep several calendars on your desktop but sync only one (perhaps named appropriately, such as "To Go" or "Mobile") with your BlackBerry.

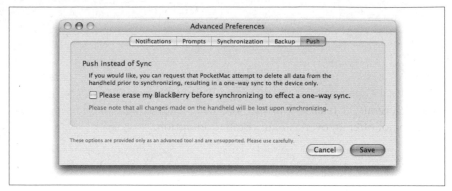

Figure 6-8. Push instead of sync

Moving on a little, you can find the most dangerous item in the PocketMac arsenal, one that allows you complete control over which notifications you get when an item is being deleted. I call this the "nuclear option" (see Figure 6-9).

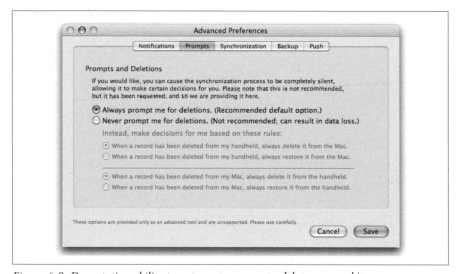

Figure 6-9. Devastating ability to auto-restore or auto-delete removed items

I can't think of a single time I'd want to activate this option unless I was absolutely certain that either my Mac or my BlackBerry was my definitive source for my data storage. I frequently add/remove items on both, so I don't think it's a good idea to manage things like this with silence. I'd rather know what is going on when things get dicey.

So not only can you sync your BlackBerry with your Mac, you also have your choice of several well-made applications to sync against. The only remaining piece of the puzzle is a Mac version of the *javaloader* binary that you can use on Windows to load applications and other packages onto the device. Hopefully the team at PocketMac will have a resolution to this soon—all signs point to this being a big deal to the PocketMac developers, especially after new handsets are released and several new bumps of firmware become available, which can only be loaded on your BlackBerry by way of a PC [Hack #20]. I was dusting off my company-issued ThinkPad to install several OS releases on my 7100 and 7290, as well as some packages that do not offer an over-the-air installation option.

While I normally wouldn't encourage people to buy an application based on future performance, I will say that PocketMac has been incredible about fixing my bugs and seems completely committed to this product. Given that as it stands right now it is the best synchronization tool for the Mac, it stands on its own for the modest price they charge for the application, and it will probably serve Mac users very well.

—R. Emory Lundberg

HACK #62 Send Voice Emails from Your Device

The power of a recorded message goes a long way. Use this hack to send voice messages via email from your BlackBerry.

There are times when you just can't get your point across in an email message. Perhaps you have an issue that needs to be conveyed with a level of insistence and importance that gets lost when sent by email. Or maybe you still haven't mastered the tiny keyboard on the BlackBerry and you have a lot to say in a short amount of time.

If you encounter any of these situations, you may find it convenient to send a recorded voice message via email. Enter pda2speak, a nifty service that allows you to do just that from any phone number (including your Black-Berry). The pda2speak service uses a web interface that is designed for small screens. You tell it the email address to which to send the voicemail, and it will call you on your BlackBerry to record a message. When you hang up, the message is sent.

Sign Up for pda2speak

Go to *http://www.pda2speak.com* in your BlackBerry Browser. The first time you use the service, you'll be asked to create an account by entering your name, email address, country, phone number, and a password. Once you're account is created, log in and you'll be taken to the main pda2speak page, as shown in Figure 6-10.

Figure 6-10. The main pda2speak page

Clicking on Send a voice e-mail will take you to a page that allows you to enter the recipient's email addresses and subject as it should appear when sent (see Figure 6-11).

Figure 6-11. Sending an important voice email

Click the CALL ME NOW button to record your message. After a few seconds, you'll receive a phone call on the phone number that you configured with your account. When you pick up the call, press the number 1 to begin recording your message. After you have completed your message, simply hang up.

What the Recipient Receives

Your voice email will be sent to the recipients you specified in surprisingly short order. The recipients you specified will receive a message with your voice message attached in a compressed *.wav* format. As shown in Figure 6-12, an additional link is given in the email to a standard *.wav* file that is stored on the pda2speak server.

You can change the phone number that pda2speak calls you at. This is useful if you want to test the service out using a landline first. Once you are comfortable using the service, you can change the number to the phone number for your BlackBerry. Pda2speak allows you to add up to three numbers for your account that you can toggle between as you send voice emails.

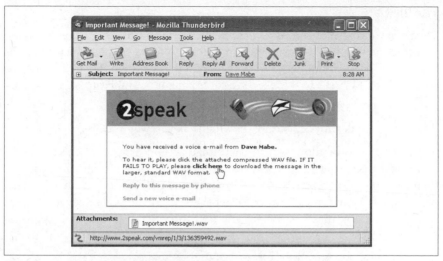

Figure 6-12. The voice email message as received by your recipients

HACK #63 Fax from Your BlackBerry

If you have access to an email-to-fax gateway, you can send your messages to the nearest fax machine—in effect, "printing" an email from your device.

Suppose you're at a client site and just received a lengthy email on your BlackBerry. You would like to have a hardcopy of the important message, but you aren't able to print from your BlackBerry. The client has several fax machines that are at your disposal, but you can't fax from a BlackBerry—or can you?

There are publicly available services that allow you to email to a fax machine. A lot of companies set up their own email-to-fax gateway services using software such as Right Fax. If you have access to such a gateway, just find the phone number of the closest fax machine and send the message by using your fax addressing scheme. Most email-to-fax gateways have an addressing scheme that is not easy to remember. These days with email as ubiquitous as it is, most people don't use outbound fax gateways frequently enough to be able to remember how to form the address when sending a fax. There is a simple program from *http://www.software-for-blackberry.com* that you can configure to remember the fax addressing scheme for your fax service so when you want to send a fax, you can simply access an option on the trackwheel menu and choose a recipient that has a fax number from your address book.

Fax the Hard Way

Here's how most fax gateways work. When you want to send a fax via email, you take the recipient's fax number and plug it into an email address that tells the fax server what number to send the email to. For example, if you'd like to send a fax to +1 (919) 555-5555, the address might be similar to:

```
/fax=9195555555@fax.server.com
```

This scheme will vary depending on which fax gateway you use.

Some fax gateways allow you to customize certain aspects of the fax using additional parameters in the address. Other parameters might include the display name or company name of the recipient for the cover sheet, a custom display name for the sender, or whether to include a fax cover sheet or not. After adding these options to the fax, the email address might become something like:

```
/fax=9195555555/com=Company_Name/from=David_Mabe@fax.server.com
```

As you can imagine, it can be a challenge to even remember the scheme at all, and if you do remember it, it can be a bear typing the address in on the BlackBerry's miniature keyboard.

Fax the Easy Way

A small program called Mail2Fax makes the difficult parts of sending a fax simple. You can download a copy of the program from Handango for $12. 95. There is no over-the-air install, so you have to install it using Desktop Manager and Application Loader.

Once installed, you access the configuration from the Home screen. The settings are shown in Figure 6-13. It includes preconfigured options for two popular fax gateways: *http://www.efax.com* and *http://www.faxaway.com*. You can also configure a custom addressing scheme. Any text you enter in the Prefix field will become the start of the address, which is immediately followed by the fax number. Text in the Suffix field goes right after the fax number. For example, if your addressing scheme is */fax=NNNNNNNNNN@fax.server.com*, the Prefix would be */fax=* and the suffix would be *@fax.server.com*.

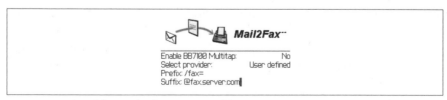

Figure 6-13. Configuring the fax gateway addressing scheme

Once you configure and save the settings, you can go to your Messages program and select the message you'd like to fax. In the trackwheel menu for any message, you'll see a Mail2Fax option, as shown in Figure 6-14.

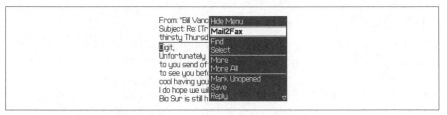

Figure 6-14. The Mail2Fax option on the trackwheel menu

Once you choose the Mail2Fax option, you are presented with a list of users from your address book, but only users that have a fax number defined. Choose the recipient or type a new fax number to use just for this fax. Mail2Fax automatically creates the appropriate email address using the rules you defined in the Mail2Fax settings and forwards the selected email message to the address. After the forwarded message is queued, you are immediately returned to the original selected mail message. This is sort of a "poor man's printer" that comes in quite handy given the right situation.

HACK #64 Sync Memos over the Air Without a BES

Keeping memos synchronized wirelessly is not just for BES users.

If you're a BlackBerry Enterprise Server user, you have access to a lot more features than your BWC counterparts. The advantages are even greater if you are on a 4.0 version of the BES. One of the best features is wireless synchronization of your server-based mailbox and all your items—mail messages, contacts, calendar, tasks, and memos. Users of the BlackBerry Web Client don't have access to this level of convenience.

There is a nice program that can get you pretty close for memos. The program is MemoPadPro and is made by J2X Technologies (*http://www.j2x.ca/*). It allows you to create memos from your device or a web interface and then have the two synchronized.

Create an Account

Create an account on the MobilePro web site (*http://www.mobilepro.net/*) to get started. Your account can be used across all the MobilePro line of applications. They have similar programs that allow you to keep track of expenses, mileage, and billing time, among other things.

You'll need to buy the program from the web site (through Handango) for $14.99. There is no over-the-air install available, so you will have to install the program using Application Loader through your USB cable.

Create Memos on the Web

The nicest feature of this program is the ability to add memos on the web site using a desktop computer or add them to the program on your device—whichever one happens to be the most convenient at the moment. The web site is well designed and lets you add new notes quickly and easily (see Figure 6-15). You can assign categories to your memos and your own custom categories. You can print your notes, or even export them to a text file from the web site.

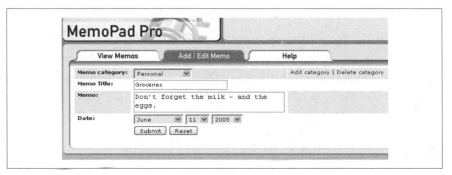

Figure 6-15. Adding a note using MobilePro.net

Create Notes on Your Device

Once you configure MemoPad Pro on your device with your username and password, you can create new memos on your device as well. Just give your note a title and some content, as shown in Figure 6-16, and use the trackwheel to choose OK from the menu. You'll be able to add a category to your note after it's created.

Figure 6-16. Adding a note using your device

Sync Your Notes

When you're ready to sync your changes, use the trackwheel to access "Sync Now" from the menu. After a brief moment, your notes will be synchronized (see Figure 6-17). The sync is bidirectional, so you'll have full access to the items you created from either location. You can add categories in either location as well. You can highlight the Date or Title header to change the order of your notes to be sorted by that field.

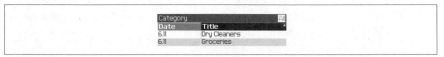

Figure 6-17. The synchronized tasks

See Also

* "Use Backpack as Your Mobile Workspace" **[Hack #47]**

HACK #65 Spellcheck Outgoing Messages

Use a free spellchecker to make up for your fat fingers, er, fat thumbs.

Many of us have become accustomed to spellchecking email messages as they are sent. Some mail program like Microsoft Outlook and Mozilla Thunderbird can be set up to automatically spellcheck messages after you hit the Send button so you'll never forget to do it. Few things can be more embarrassing in the world of email than misspelling a word in some silly way—it comes across as sloppy and unprofessional, especially in light of the fact that it is so easily avoided.

It can be an adjustment for new BlackBerry users who are used to relying on the spellchecker in their desktop email program. After all, there is no built-in spellcheck solution for the BlackBerry yet, and you're typing with your thumbs for goodness' sake!

Some software vendors aren't waiting for a solution from RIM and they're capitalizing on this need. One such vendor is a small company call SomeDevelopers. Their BlackBerry spellcheck solution called bbSpell has a similar pricing scheme as the Opera **[Hack #89]** web browser—you can use it for free for five spellchecked messages a day if you are willing to view some advertisements when you use the product. For an annual fee of $39.95, you can use the product advertisement-free with unlimited spellchecks. Although you can't configure bbSpell to check messages automatically, you can invoke it manually to prevent a potentially embarrassing spelling gaffe.

Install bbSpell

To install bbSpell, you'll need to pony up a little personal information on an online form including your name, email address (they won't share it), a phone number, and whether your company pays for your BlackBerry or not.

Once the form is completed, an email is sent to the address you specified explaining how to access and download the over-the-air install. In case you aren't able to install an over-the-air download (some companies disallow this using IT policies), the email includes the files required to install bbSpell with Application Loader. Once installed, bbSpell places a professional-looking icon on your Home screen.

Spellcheck a Message

To spellcheck a message, compose a message and then use the trackwheel to access the menu. Choose the Check Spelling option, as shown in Figure 6-18.

Figure 6-18. The Check Spelling option on the menu

At this point, bbSpell communicates with its web service component to send the subject and body of your message to a server that does the heavy lifting of the actual spellchecking. As it communicates with the server, the unregistered version displays a brief advertisement. Figure 6-19 shows the results as they are returned. Notice there the bold text that indicates the unrecognized words that I misspelled on purpose (no, seriously, I did misspell these on purpose).

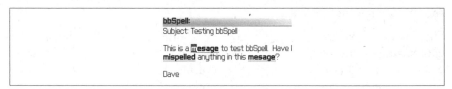

Figure 6-19. Spellcheck results

You can use the trackwheel to scroll between misspelled words. When selecting a word, click the trackwheel to bring up a menu (see Figure 6-20) that has suggestions and options for adding the word to your user dictionary (which is stored locally on your device) or changing the word to something else.

> If you misspell a word in the same way more than once in a message, choosing a correction that bbSpell suggests corrects only the selected instance of the word, not all instances like it probably should. You can correct all instances of the misspelling at once by selecting Change and typing the correction.

Figure 6-20. The bbSpell menu

Once you are finished spellchecking the message, use the trackwheel to choose Send from the menu. When you are using the unregistered version of the product, an advertisement is displayed on the screen for five seconds, as shown in Figure 6-21. At the end of five seconds, you can use the trackwheel to visit the advertised site or choose Close to return to your message list.

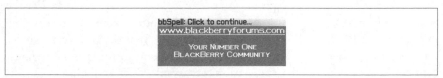

Figure 6-21. An advertisement in the unregistered version of bbSpell

For companies that would like to use bbSpell but are concerned about message content being delivered to a third party over the Internet, the server-side bbSpell component can be purchased and installed on a company's server. The custom URL that tells the client component which server to connect to for the spellcheck can be sent to all users at once using an IT policy [Hack #99]. There are options for SSL, forcing connections to BES/MDS, and using a server-side site license that can also be pushed out by using a BES and IT policies.

HACK #66 Try a Third-Party Web Browser

This alternative web browser is nice and quick; it also gives BWC users access to full HTML content and more.

If your company has a BlackBerry Enterprise Server with MDS enabled, then you've probably already got access to an excellent HTML web browser. If you're a BlackBerry Web Client user, then hopefully you're using the Black-Berry Internet Browser. Either way, this third-party browser has some pretty amazing features that are worth your consideration.

The Webviewer browser (provided by Reqwireless) has several features that set it apart from the built-in BlackBerry Browser. Not only can you read several more formats than with the built-in browser, but many users consider it to be much faster.

Install and Use WebViewer

Just like most current third-party applications, you can install WebViewer over the air (*http://www.reqwireless.com/wap_bb*) or download the archive suitable for installation through Application Loader at *http://www.reqwireless.com/download-webviewer-bb.html*. You will need to have configured your device for TCP/IP access for the browser to work [Hack #37].

WebViewer works much like the BlackBerry Browser does; it communicates directly to a gateway that proxies the HTTP requests. The gateway works in tandem with the browser to format the content optimally for your device. This framework is almost identical to the Mobile Data Service that comes with the BlackBerry Enterprise Server.

Once installed, it works much as you'd expect a browser to work. On the main screen that appears after running the application, click the trackwheel and choose URL from the menu. Type the URL that you'd like to go to, click the trackwheel, and choose OK from the menu to initiate the request. Notice the excellent feedback given by the progress bar in Figure 6-22; this is one feature that sets it apart from the built-in BlackBerry Browser.

Figure 6-22. Loading a page in WebViewer

Unique Features

There are some other excellent features in WebViewer that the BlackBerry Browser has yet to match. WebViewer allows you to view individual cookies for particular domains. You are able to delete individual cookies, delete all the cookies for a particular domain, or delete all your cookies at once (see Figures 6-23 and 6-24). The BlackBerry Browser won't let you view *any* cookies, and if you want to delete one of them, you have to delete them all.

Not only can you view several file formats that the BlackBerry Browser cannot handle (Microsoft Word, Excel, PDF, WordPerfect, etc.), but Web-Viewer will also let you try to attempt to use Optical Character Recognition (OCR) on TIFF files and other images. Select any image in the browser and click the trackwheel. Choose OCR Image from the menu and WebViewer

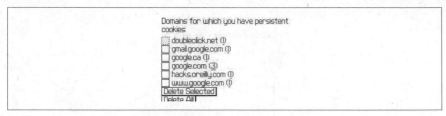

Figure 6-23. Cookies by domain

Figure 6-24. Viewing and deleting an individual cookie

will make an attempt to "read" the text in the image. For example, it was able to read the O'Reilly logo in GIF format on the O'Reilly home page (*http://www.oreilly.com*), as shown in Figure 6-25. Although the OCR function isn't perfect, some readers might find a use for this feature.

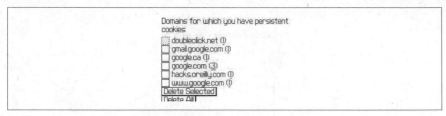

Figure 6-25. The OCR function in WebViewer

WebViewer also has had JavaScript support long before the BlackBerry Browser supported it. It also has nice support for frames. When the Black-Berry Browser encounters a page that uses frames, it displays a page that asks you which frame you'd like to display. Once you view one of the frames, the only way to view a different one is to go back using the Escape button—not exactly an optimal user experience. WebViewer just goes ahead and displays the frames inline on the same page. Figures 6-26 and 6-27 show a page with frames in the BlackBerry Browser versus WebViewer. Notice with the BlackBerry Browser you can only view a single frame at a time, while WebViewer displays pages with frames.

Figure 6-26. Frames in the BlackBerry Browser

Figure 6-27. The same page in WebViewer

Try Another Email Program

HACK #67

This third-party app boasts some nice features and lets you access mailboxes with POP or IMAP.

If you are a BlackBerry Enterprise Server user, then your administrator may restrict certain functions on your BlackBerry. One of the first features that a company may choose to restrict is the ability for your device to use multiple *service books*.

A service book enables your device to receive mail from a particular service, such as your BlackBerry Enterprise Server or the BlackBerry Web Client. If your company's IT policy restricts access to the BlackBerry Web Client, you may not be able to access your personal email accounts with the BlackBerry's built-in Mail application.

If your secondary mail account is accessible through a web interface, you may try using the BlackBerry Browser to access it. If you can't, you're not out of luck yet. Try this product from ReqWireless called EmailViewer. It is a full-featured email client that can access your email by using POP or IMAP.

Install and Use EmailViewer

Most third-party software packages offer over-the-air installs in addition to downloads for use with Application Loader. EmailViewer is no different and, like most, the over-the-air install is far more convenient.

Once installed, you need to set up your email accounts in the program. The navigation in EmailViewer is certainly not optimal. Most actions require a trackwheel click to access the menu, which takes a little getting used to. To set up your email account, scroll to Accounts on the main EmailViewer screen, click the trackwheel, and choose Select from the menu, as shown in Figure 6-28. This brings you to the Accounts list. Click the trackwheel again and choose New to set up a new account.

Figure 6-28. The main EmailViewer screen

In the new account configuration screen, you'll find the familiar options for setting up a typical email client. Notice there is support for Hotmail, Secure POP3, and Secure IMAP, in addition to the standard POP3 and IMAP protocols (see Figure 6-29). There is an Autofill option available on the menu to populate these settings automatically for various popular email providers like AOL, BellSouth, etc.

Figure 6-29. New account configuration

So where is the configuration for setting your outgoing mail via SMTP? It is there, although hidden. By default, the SMTP settings are blank, which tells the client to use ReqWireless's mail relay settings. If you'd like to use another SMTP relay, click the trackwheel and choose SMTP from the menu.

> Note that most SMTP relay hosts require clients to be in a certain IP network range to send outbound email. Since you'll be going through the ReqWireless proxy, the IP address that your SMTP server will think you're coming from will likely not be in its allowable IP range (it better not be!). Most SMTP servers support authenticated SMTP, which requires you to enter a username and password to override the IP range restriction, although there are no guarantees that this feature is enabled on yours.

Send and Receive Messages

Once you've configured your accounts, return to the main EmailViewer screen and select Receive. If you've entered more than one account, you will be prompted to choose which account you'd like to view. Figure 6-30 shows the view of an IMAP mailbox in EmailViewer. You get full access to subfolders along with the most recent messages you've received in your account. The last 10 messages will be displayed, although this setting can be changed. You can also choose whether to display the subject and sender in the message list.

Figure 6-30. Viewing a mailbox via IMAP in EmailViewer

The process of composing a message is much different than in the built-in Mail program and will at first feel a little clunky. Instead of showing the To, Subject, and Body fields on a single screen, EmailViewer takes you through each field, one screen at a time, requiring trackwheel clicks to progress. Once you send your first message (see Figure 6-31), the next one will come with ease. Fortunately, EmailViewer integrates with your BlackBerry Address Book so you can use those email addresses when composing new mail. There is also an interface for adding new contacts from within Email-Viewer that automatically get added to your device's Address Book.

The first time you send a message through the ReqWireless relay, you will need to verify your identity: ReqWireless will send a one-time authentication email to that email address before it allows you to send through its relay.

Figure 6-31. The last step of the compose process

Other Features

EmailViewer also boasts some nice features not found in your BlackBerry Mail application. Not only can you can view images that are sent with your mail messages, but you can view some Office documents that are sent as attachments. You can also turn on a spam filter, which hides messages it believes to be spam from your message list (without actually deleting them from your account). EmailViewer gives you the ability to send plain text email or HTML email and you can change the reply-to address on outgoing messages. It also lets you configure a signature that is appended to all outgoing messages.

HACK

#68

Edit Office Documents

This software from Dynoplex actually lets you modify Microsoft Word and Excel documents right on your device.

RIM has made great strides in unlocking message attachments for viewing on the handheld. The Attachment service when released with BES 3.5 was a big hit, and it's been improved ever since. The attachment viewing functionality has even been extended to BlackBerry Web Client users. However, there is no way to make modifications to those attachments using the native BlackBerry software on the handheld.

Dynoplex has filled this gap by offering a solution called eOffice. There is a client component that works in conjunction with a web service that allows you to save, edit, and resend Excel and Word documents right from your Black-Berry. The Professional version even comes with a PIM application that lets you view your messages in a preview pane. Dynoplex also sells licenses for their server software that can be installed in your own data center so your documents can be processed without data traveling over the public Internet.

eOffice Versions

Dynoplex eOffice can be found at *http://www.dynoplex.com*. There are currently three different versions of eOffice, each with increasing levels of functionality. The Basic version ($119.95) lets you create new Word and Excel documents. With the Standard version ($149.95), you can save and edit the Word and Excel documents that come attached to your inbound email messages. The Professional version ($199.95) provides all that the Standard version does, plus a spellchecker, a PIM program, and the ability to send HTML-formatted email messages.

There is a 30-day free trial available for eOffice from *http://www.dynoplex.com/downloads.shtml*. There is no over-the-air install, so you'll have to use Application Loader to complete the install. The installation program is more like a Windows program than the typical *.alx* and *.cod* files that you're probably used to. The 5-MB installation file will install some documentation on your computer, and then automatically bring up Application Loader to install the appropriate files on your device.

Use eOffice

Once you have eOffice installed, you can start creating new documents by accessing eOffice from the Home screen. Once you're in eOffice, you can choose eWord to create Word documents or eCell to create Excel spreadsheets (see Figures 6-32 and 6-33).

Figure 6-32. eWord

Figure 6-33. eCell

When you are ready to save your document, you can choose to save it locally on the flash memory of your device, or you can store it on the Dynoplex service using the storage allocated for your account. To send your document via email, open the document in eOffice and choose Email from the File menu. Choose a contact to send to or use a one-time email address for the recipient. You can email documents that you've stored on your device or ones you've stored on Dynoplex's service. The recipient will receive the attachment as if you emailed it from your computer.

Some of the more complex formatting in Office documents gets lost in the conversion to eOffice formatting. For example, Word tables get reduced to flat lists.

Edit and Save Attachments

If you've purchased the Standard or Professional version, you can save the attachments of incoming messages to eOffice. You'll see an additional item on the trackwheel menu when viewing a message, as shown in Figure 6-34.

Figure 6-34. The Save to eOffice menu item

When you choose Save to eOffice, the message is actually forwarded to Dynoplex's service, converted, and returned to your device in a format that is viewable with the eOffice client software.

The eWorks

The Professional version of eOffice comes with a program called eWorks that lets you view your data in a way that is similar to the folder list on your Microsoft Outlook client (see Figure 6-35).

Figure 6-35. eWorks on your device

This lets you work with the data on your BlackBerry using a different view. Some users might like this.

Display a Slideshow

Use this program to cycle through your photographs on your BlackBerry.

The built-in Pictures program that was added to the BlackBerry in Version 4.0 is a nice feature. One limitation of the program is to display your own pictures, you must use the BlackBerry Browser to add them. Because the browser is only able to view web sites, you have to post your photos on a web site to add them to your device. This is an obstacle that, although not too difficult or costly, is certainly inconvenient. After all, you've got a device that's optimized for email—why can't you simply attach your photo to an email to yourself?

It turns out there is a third-party program that allows you to do just that. Berry Pix from Colabnet, Inc. allows you to add custom photos via email. Not only that, but Berry Pix also allows you to display a slideshow of your photos as a replacement for your device's standby screen. This functions much as a screensaver on your computer would.

Install Berry Pix

You can download Berry Pix from its web site at *http://www.colabnet.com/berrypix.html*. There is no over-the-air install available, so you'll have to use the Application Loader download from the web site and install it with the BlackBerry Desktop Manager. There is a seven-day trial version available.

There are two versions of Berry Pix: one for devices with over 8 MB of memory, and another for devices with less than 8 MB of memory. Both of these versions are included with the Berry Pix download. To see how much memory your device has, go to Options → Status from your Home screen and then check the File Total value. If your device has less than 8 MB, choose the *Berry Pix Small Memory.alx* file when you install with Application Loader; otherwise, choose the *Berry Pix.alx* file

Add Images to Your Device

After installing Berry Pix, you'll be able to add images to your device by sending an email. Select the image you'd like to make available on your device. To add it to your BlackBerry, make a copy of it and rename the copy to have a prefix of *x-rimdevice*. For example, a photo with a filename of *sunflower.jpg* would become *x-rimdevicesunflower.jpg*.

> Why do you have to rename the image to have a prefix of *x-rimdevice*? Whether you are a BES user or a BWC user, an attachment will not be forwarded to your device—only a reference to the attachment. Only when an attachment is sent with the *x-rimdevice* prefix will the entire attachment be sent to the device.

Once the email is received on your device, scroll to the attachment portion of the message and click the trackwheel to bring up the menu. Choose Add to Berry Pix from the menu, as shown in Figure 6-36.

Figure 6-36. The Add to Berry Pix option on the menu

This loads your image file into the local storage on your device and displays the Berry Pix Manage Your Pictures screen, as shown in Figure 6-37. It tells you how much space the picture is consuming on your device and how many more pictures you have room for.

Figure 6-37. Your picture added to the device

Let the Slide Show Begin

After you've installed Berry Pix, it will act much as a screensaver does on your desktop computer. After five minutes of inactivity, a slideshow will appear on your device with all the pictures on your device. Berry Pix displays a similar set of features as PocketDay [Hack #60]—it acts much the same as the normal standby screen appears on most devices (except the 7100 series). Figure 6-38 shows the Berry Pix screen along with the slideshow that appears.

Figure 6-38. The Berry Pix home screen

The trained eye will notice an ever so slight difference (although 7100 users will notice a huge difference) between the icons that appear on the Berry Pix home screen and those that appear on your regular Home screen. Berry Pix is, in effect, emulating the BlackBerry Home screen of non-7100 series devices, attempting to make it appear as close as possible to your actual Home screen.

Make Adjustments

You can configure a few Berry Pix options to control the behavior of the slide-show. You can adjust the inactivity timeout from the default of five minutes to an amount that you prefer. Also, you can change the amount of time each picture is displayed in the slideshow from the default of 10 seconds.

You can also assign a category to each of the pictures you've stored in Berry Pix and then instruct the slideshow to display only pictures of a certain category. From the Manage Your Pictures screen in Berry Pix, use the track-wheel to bring up the menu. Choose the Add to Category option from the menu. Figure 6-39 shows the screen that appears, which allows you to assign a new category to the selected picture.

Figure 6-39. Creating a new category for your picture

HACK #70 Get Real-Time Stock Quotes and Charts

Stay in tune with the markets while you're on the road to satisfy the inner day trader in you.

If you are like a lot of people, you would trade in the stock market if you only had the time. Trading stocks on a long-term basis is difficult, but trading on a short-term basis requires discipline, a time commitment, and a watchful eye on the markets. If you spend a lot of your work week traveling, you still need to keep a close watch on your portfolio or it will suffer.

You can use this software package from Quotestream to get access to real-time stock quotes and charts right from your BlackBerry. Also, most major online brokerages provide WAP interfaces to trade, and Quotestream gives you easy access to your online broker right from the program.

Quotestream

Quotestream is made by QuoteMedia, Inc., a stock data service company. As is common with companies built on providing stock data, you'll be able to download the software for free, but it won't be functional until you get an account from QuoteMedia. The pricing for the accounts can be found by clicking on the Register link on the Quotestream web site (*http://www.quotestream.com*). The account includes access to the Quotestream desktop software as well.

The Quotestream wireless software is available only as an over-the-air download. Go to the URL (*http://app.quotemedia.com/jsp/quotestream/wireless.jsp*) for instructions on retrieving the download. The site will ask for the version of BlackBerry, you have and then you have the option of sending an email to your device that contains the URL for the download or using the WAP interface that you can go to directly in your BlackBerry Browser to download and install the software.

Use Quotestream

Once your account is set up, you can use the software on your BlackBerry. Figure 6-40 shows the screen you'll see after you log in. You'll have to get used to clicking with the trackwheel instead of simply using the Enter key as a shortcut for most functions in this program. Choose Streaming Portfolios to access the streaming real-time quotes. You can maintain up to five different portfolios with up to 45 symbols each. Figure 6-41 shows the streaming quotes for your portfolio of ticker symbols.

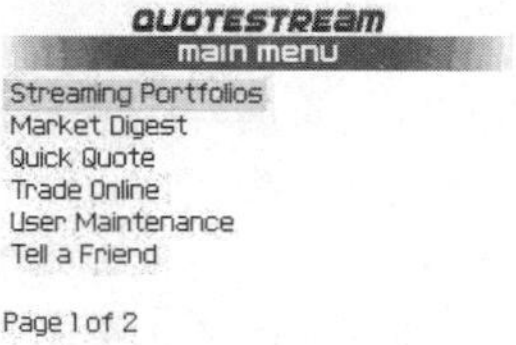

Figure 6-40. The Quotestream home screen

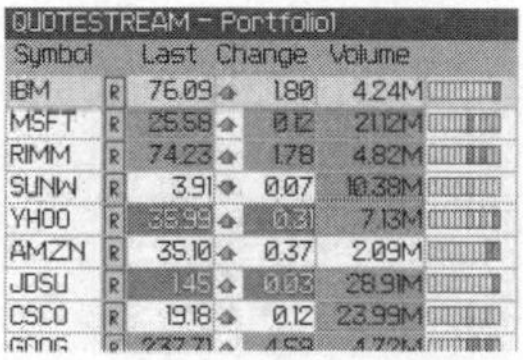

Figure 6-41. Quotestream's portfolio view

To modify your portfolios or create a new one, click on the trackwheel and choose Edit Portfolio. Figure 6-42 shows the Edit Portfolio screen that appears. The interface to modify symbols is easy to use—superior to other products on the market.

Real-Time Stock Charts

To access real-time stock charts for any of the symbols in your portfolios, use the trackwheel to select the symbol by highlighting it. Click the trackwheel to

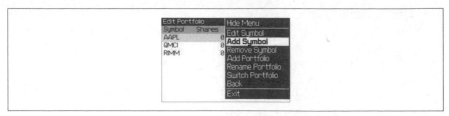

Figure 6-42. Editing a portfolio in Quotestream

bring up the menu, and select Charts. This takes you to a screen similar to the one in Figure 6-43.

Figure 6-43. A daily stock chart for RIMM

If you're an avid trader, you'll notice the familiar chart design of price on the top and volume across the bottom. You can configure the time interval for the chart anywhere from intraday to a five-year chart. You can customize the type of chart that is displayed as well. You can choose from bar charts, line charts, area charts, and the ever popular Japanese Candlestick charts.

Level II Data

You can view Level II screens on your device as well. Level II quotes is a service that allows you to view the bid and ask prices of individual market makers so you can see the "market" as it happens. Figure 6-44 shows the depth chart for a security in our portfolio.

Depth – RIMM (RT)					
MMID	Bid	Size	MMID	Ask	Size
BRUT	74.230	11	SIZE	74.250	2
SIZE	74.230	1	BRUT	74.270	1
LEHM	74.180	1	LEHM	74.340	1
SCHB	73.980	2	RBCM	74.350	1
COWN	73.900	1	NFSC	74.360	1
TMBR	73.810	4	BEST	74.390	1
NITE	73.790	1	TMBR	74.390	1
PRUS	73.650	1	SCHB	74.400	10

Figure 6-44. Level II depth chart for RIMM

Set Alerts

You can set alerts for maximum and minimum price and volume for particular stocks in your portfolio. When the criteria are met, you can configure your device to notify you automatically. Quotestream also provides convenient links to the WAP interfaces of brokers that have them by clicking the trackwheel and choosing the Trade Online option from the menu. You no longer have to use "business travel" as an excuse to stay out of the virtual trading pit!

Modify on Desktop, View on Handheld

One of the best features of Quotestream is the integration with the desktop version of the software. You can access their real-time quoting service in your desktop browser and modify your portfolios there when you have the convenience of a full-size keyboard. When you log into Quotestream on your BlackBerry later, you'll see all the changes that you made to your portfolios on your desktop computer.

> Quotestream conveniently lets you log in to more than one location simultaneously (for example, your desktop and your BlackBerry). However, the location you logged into most recently will have real-time quotes, and all the other locations immediately revert to quotes that are 15 to 20 minutes delayed.

See Also

- *Online Investing Hacks* (O'Reilly, 2004)
- QuoTrek, *http://www.quotrek.com*

HACK #71 Track Your Fitness

Use your BlackBerry to monitor your progress on your quest for Olympic glory, or just your quest to lose a few pounds.

As a running geek and also a technology geek, it's nice when the two worlds meld together. I've kept track of the miles I've run for years, but I've never been completely satisfied with putting all my entries into a paper-based log. With all the advances in technology over the past several years, I've still been clinging to my running log in book form like a toddler who's a little too old for a pacifier.

The problem with computer-based exercise logs is that you have to be in front of the computer to enter data into them, and it's not practical to bring your computer along with you when you go to exercise—unless that computer is a BlackBerry!

Use a Running Log

A company called b4 Technology makes a nice running log with all the features I need. Both a four-day trial version and the full version of Running Log for BlackBerry are available at *http://www.b4technology.com/RunningLog.html*. When you run the program for the first time, you'll be asked to enter some personal information (see Figure 6-45), including age, height, gender, and weight to be able to calculate calories burned when you enter your running entries.

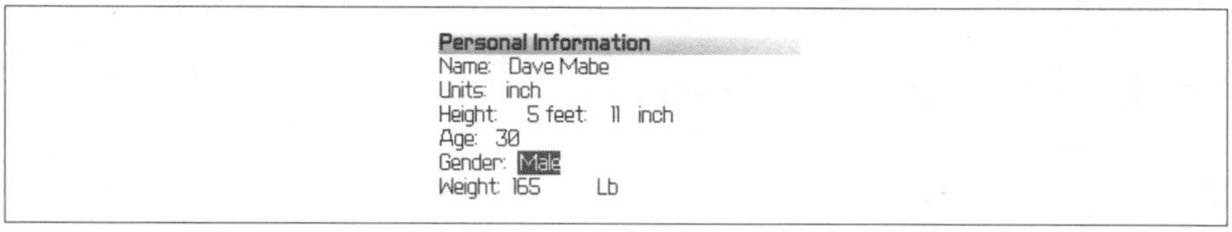

Personal Information
Name: Dave Mabe
Units: inch
Height: 5 feet   11 inch
Age: 30
Gender: Male
Weight: 165   Lb

Figure 6-45. Entering personal information into the Running Log

After saving your personal information, select the log icon on the Running Log main screen and click Enter to start adding entries to your log. Figure 6-46 shows the New Entry screen. Entering the Duration and the Distance of your run automatically populates the Pace and Calories fields. The Weight and Average Heart Rate fields are optional. One nifty feature of the log is its use of custom *courses*. As you add your log entries, you can add the course that you ran and give it a name. The next time you add an entry, you can choose from the list of all courses that you've added. This is nice to compare your marks with the previous times you've run the same course.

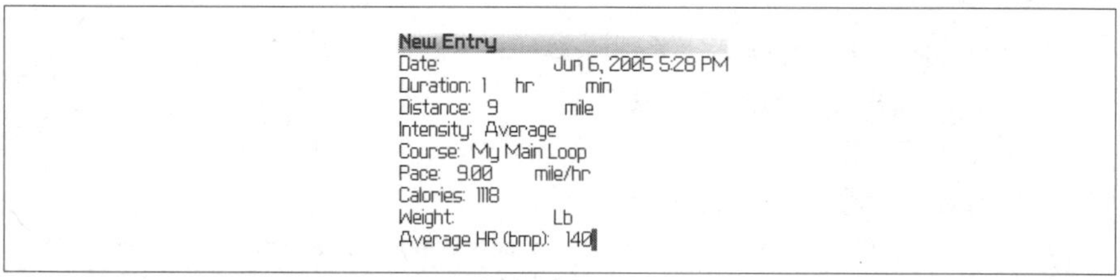

New Entry
Date: Jun 6, 2005 5:28 PM
Duration: 1   hr   min
Distance: 9   mile
Intensity: Average
Course: My Main Loop
Pace: 9.00   mile/hr
Calories: 1118
Weight: Lb
Average HR (bmp): 140

Figure 6-46. Adding a new entry to the log

As any runner who has ever kept a log knows, the weekly total mileage is the most popular gauge of training. This running log program provides nicely formatted graphs for weekly and monthly mileage and duration. See Figure 6-47 for an example of a weekly duration graph.

Track Your Caloric Intake Versus Exercise

For the ultra–Type A personalities out there, b4 Technology also provides a product that allows you to track your caloric intake and compare it against the calories you burn exercising. You can set weight loss goals and monitor

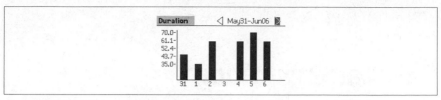

Figure 6-47. Weekly running by minutes graph

your progress at attaining them. The product is called Total Fitness for BlackBerry and can be downloaded from *http://www.b4technology.com/ TotalFitness.html*. Like the Running Log for the BlackBerry, there is a four-day free trial available. The full version is available for $29.99.

Total Fitness also asks you for some personal information to customize the product. You can set weight loss goals and follow a diet plan. You can enter your daily meals (see Figure 6-48) and compare them to the diet you've chosen to follow. Total Fitness contains a large database of foods to choose from. I found my Cheerios, in there and also the triple cheeseburger I ate at lunch.

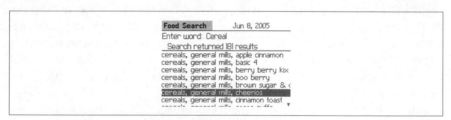

Figure 6-48. Searching Total Fitness's food database

You can enter your exercise and it computes the calories burned. It contains a list of different types of exercises (cycling, running, table tennis, etc.) and the amount of calories the activity typically uses. Figure 6-49 shows the list of workouts you can choose from. You can also add your own to the list (wood chopping doesn't appear by default).

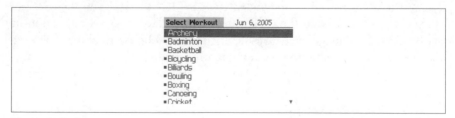

Figure 6-49. Exercise activities

It uses your personal information to calculate the calories burned during each activity. You can display a variety of graphs and reports that munge the data you've entered in various ways.

CHAPTER SEVEN

BES Administration

Hacks 72–84

It's fairly easy to get a BlackBerry Enterprise Server off the ground. However, as many a BlackBerry administrator will tell you, it is a difficult and time-consuming chore to keep the service running smoothly. For the most part, it is no fault of RIM's—there are many tools provided that continue to go unused in many BlackBerry shops. There are more than a few tips and tricks to keep the BlackBerry server and your users happy. The motivated BlackBerry administrator will find several gems in this chapter from simply adding several users in one fell swoop [Hack #72] to implementing security [Hack #73]. There are several hacks you can use to send yourself proactive alerts when problems arise—just be careful when sending fault alerts to your BlackBerry, since your service may be down (that's the whole point, right?). Of course, there is no substitute for a managed service—there are several companies who will manage your BlackBerry infrastructure for you.

HACK #72 Add Users to the BES in Bulk

Use a specially formatted file to add multiple users to your BlackBerry Enterprise Server in one fell swoop, saving you lots of pointing and clicking.

When a new user needs to be added to your BlackBerry Enterprise Server, normally you would add the user using the standard Add User function in the BlackBerry Management Console. You would choose to add a new user and then pick her name from your company's address list. For one user at a time, this process works just fine. Would it still be as easy if your company just bought another company and you had to add 50 new users to your BES? Of course not—that is far too many points and clicks. As BlackBerry users, we like to find efficient ways to accomplish tasks.

There is a way to add multiple users to your BlackBerry Enterprise Server all at once instead of one at a time. It requires a little work ahead of time.

Generate the Mailbox File

You'll need to create a file that contains the X.500 directory names for the mailboxes of the new users. In a Windows Active Directory domain, this corresponds to the *legacyExchangeDN* property on a mailbox-enabled user object. An example would be:

```
/o=org/ou=site/cn=container/cn=alias
```

If you use Microsoft Exchange for your email platform and you are running a Windows 2000 Active Directory domain, the best way to retrieve these values in bulk is with the *csvde.exe* utility that is installed on any Windows 2000 Server. Go to the command prompt on the server and type the following command to export all users in the domain along with their X.500 directory names:

```
csvde -f file.csv -d dc=yourdomain,dc=com -l DN,legacyExchangeDN -p subtree
```

This will create a comma-delimited file called *file.csv* suitable for opening in Microsoft Excel in the current working directory that contains the *legacyExchangeDNs*. Add the appropriate *legacyExchangeDN* values for each user to the file you created with one value per line, save the file, and, if needed, copy it to the machine on which you run the BlackBerry Management console.

Import the File

In your BlackBerry Management Console, right-click on the server to which you'd like to add the new users, then go to Add Users → Import Users from File, as shown in Figure 7-1.

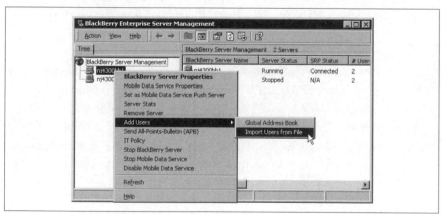

Figure 7-1. The Import Users from File option

Point to the file you just created and click Open. A dialog box will appear asking if you would like statistics to be cleared if any exist in the mailbox. If any of these mailboxes have been BlackBerry enabled in the past, you may want to choose Yes here to continue with the previous BlackBerry counters for the particular mailbox instead of resetting them—otherwise, choose No.

> The user statistics prompt is always one that confuses administrators who are new to BlackBerry. The BlackBerry Enterprise Server keeps counters of messages sent, received, and filtered, along with a last contact timestamp. Except for the last contact time, which resets every time the user receives an email anyway, the user statistics are largely meaningless, since the date the user was added to the server isn't kept. Even if you find value in these statistics, it's hardly worth confronting the administrator with an ominous Yes or No dialog box every time a user is added to the BES.

All the users in the file will be added to your BES. If there are any names in your file that cannot be resolved to a mailbox, you will be conveniently prompted to correct them as the import occurs.

Lock Down Your BlackBerry

HACK #73

With all the sensitive information kept on handhelds, no wonder security organizations cringe as more users obtain BlackBerry devices and PDAs. Some simple steps can help put everyone's minds at ease.

If lost or stolen, handhelds contain proprietary information that could be compromised. There are some features available that individual users can take advantage of. Additionally, there are options available to BlackBerry administrators to help ensure that corporate information is protected.

Set Owner Information

If you lost your wallet, you'd want it to have your contact information in case the finder wants to return it to you. Likewise, you would want your contact information contained in your handheld for the same reason. Owner information is one of the first things that should be set up on your handheld.

Owner information will be displayed when the BlackBerry is locked and password protected. At a minimum, name and contact phone number or address should be included. To set owner information on your device, open up your handheld options (on 7100 series devices, it's under the Settings option). Click Owner and type your contact information. Click the trackwheel and select Save. The owner information will appear when your handheld is locked, as shown in Figure 7-2. This can also be automated using BESAdminClient [Hack #83].

Figure 7-2. Owner information displaying when the handheld is locked

Set Up a Password and Security Timeout

If your handheld contains sensitive information, you can set a password and security timeout on your handheld to protect your data. An individual password between 4 and 14 characters can be set. In the handheld options, click Security. Set the Password field to Enabled. Security Timeout can also be set to various values ranging from one minute to one hour. The Security timeout defines the period of inactivity after which the handheld will be locked and require a password to unlock. Once you have set these, click and select Save. You will be prompted to enter a password and then asked to enter the same password again to confirm. Security options are displayed here in Figure 7-3.

Figure 7-3. Handheld security options

Lock a BlackBerry Remotely

If you're an administrator, your BlackBerry Management console provides options useful if a handheld is lost or stolen. Using IT Admin, shown in Figure 7-4, choose Set Password and Lock, Set Owner Info, or Kill Handheld.

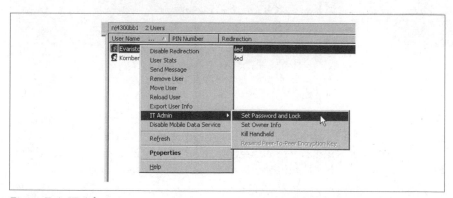

Figure 7-4. IT Admin options

IT Admin commands are sent over the air and therefore the handheld must be able to receive the commands!

To access the options:

1. Open BlackBerry Enterprise Server Management. In the left pane, your BlackBerry server will be displayed.

2. Select a server name. Users assigned to that server are displayed in the right pane.

3. Right-click a user and select IT Admin.

Set Password and Lock, shown in Figure 7-5, can be used to create a new handheld password over the air and lock the handheld. It will replace any existing password on the handheld. This action will lock the handheld and prompt the user for the new password. If a password exists on the device, the user will be asked to accept the new password that the administrator set.

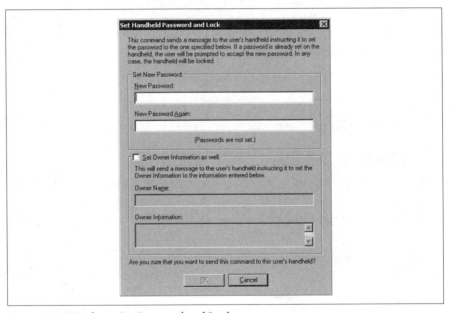

Figure 7-5. IT Admin, Set Password and Lock

Set Owner Info can be used to set the owner information that is stored on the handheld. Kill Handheld is another option for a lost device—it will disable the handheld and delete all information that was stored on the handheld. The Kill Handheld confirmation screen is displayed in Figure 7-6.

If a handheld is disabled using Kill Handheld, it can be reenabled at a later time by running the BlackBerry Application Loader on the BlackBerry Desktop Manager. The data will be wiped from the device unless you restore it from a backup.

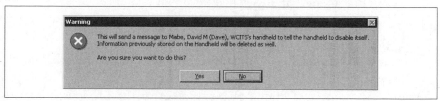

Figure 7-6. IT Admin → Kill Handheld

—Shari Kornberg

Test Your SRP Connection
HACK #74

Use this utility to verify your BES's connection to RIM's network.

The SRP connection between the BlackBerry Enterprise Server and Research In Motion's network is critical. If this connection goes down for any reason, all BlackBerry devices homed on that BES will not be able to send or receive emails. This makes the SRP connection vitally important to the service.

How do you know if the SRP connection is down? RIM provides a convenient utility to run from the BlackBerry Enterprise Server to allow you to test the SRP connection at any time from a command prompt. It makes a connection with the SRP node on RIM's network that you have configured for each BES on your server machine.

Run bbsrptest.exe

You'll find the *bbsrptest.exe* utility in the following directory within the BlackBerry Enterprise Server installation:

C:\Program Files\Research In Motion\BlackBerry Enterprise Server\Utility

To run the tool, bring up a command prompt and *cd* to the directory. Run the utility by typing **bbsrptest** from the command prompt without any options. If your SRP connection test is successful, you'll receive output similar to the following:

```
NetworkAccessNode is srp.na.blackberry.net.
Attempting to connect to srp.na.blackberry.net (206.51.26.33), port 3101
Sending test packet
Waiting for response
```

```
Receiving response
Checking response
Successful
```

Because you can configure more than one BES per server and each BES can be configured to point to a different SRP node on RIM's network, you'll get the previous output for each BES you've set up on your machine.

If the test fails, you will receive output similar to the following:

```
NetworkAccessNode is srp.na.blackberry.net.
Attempting to connect to srp.na.blackberry.net (192.168.0.222), port 3101
connect( ) failed: Connection timed out (10060)
```

Notice the IP address of srp.na.blackberry.net in the previous output. I used the local *HOST* file to point srp.na.blackberry.net to a nonexistent address on my local network so I could generate this error for you to see.

Run bbsrptest.exe from Your Workstation

When you run the *bbsrptest.exe* utility from a computer without the Black-Berry Enterprise Server installed, you'll get the following output:

```
Registry key HKEY_LOCAL_MACHINE\Software\Research In Motion\BlackBerryRouter is
missing, trying HKEY_LOCAL_MACHINE\Software\Research In Motion\BlackBerry
Enterprise Server\Dispatcher
Registry key HKEY_LOCAL_MACHINE\Software\Research In Motion\BlackBerry
Enterprise Server\Dispatcher is missing
```

This is because the tool needs to determine which SRP node your BES instances are configured to connect to. If the BES software is not installed, there is no way for the tool to know how to connect to RIM's network to verify connectivity.

It turns out that the registry key and its contents are all the tool needs. You can export the registry key from the server and import it on your workstation, and you can run the utility from there.

Export the registry key using regedit. Go to the following key:

```
HKEY_LOCALMACHINE\Software\Research In Motion\BlackBerry Enterprise Server
```

Right-click on the previous key and choose Export from the menu, as shown in Figure 7-7.

Save the exported key to a file and then copy it to the computer from which you'd like to be able to run *bbsrptest.exe*. Import it into the local registry by double-clicking on the *.reg* file. Then simply copy the *bbsrptest.exe* file to a local directory and you'll be able to run the utility from the computer without BES software installed.

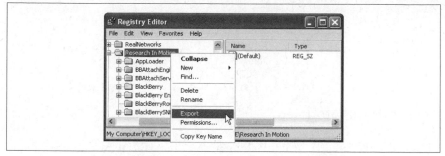

Figure 7-7. Exporting the registry key

 A successful run of *bbsrptest.exe* from your local workstation tells you that your machine was able to connect through
port 3101 to RIM's network. This doesn't necessarily mean
your server computer will be able to do the same—for a true
test, run this from your BES server.

Send an Alert When Your SRP Is Down

HACK #75

Wrap some simple Perl code around bbsrptest.exe to create an automated
alert when connectivity is lost.

You already realize the importance of you SRP connection. If connectivity is
lost, all your devices are unable to send or receive email, view web sites via the
BlackBerry Browser, or view attachments to email messages they've already
received. If there are more than a few heavy BlackBerry users on your Black-
Berry Enterprise Server, it won't take long for your phone to start ringing.

As with any service outage, it always makes your job a little easier when you
already know about and are working on a problem when the CEO calls and
asks about it. Use this hack to send an automated alert when *bbsrptest* fails.

The Code

```perl
use Mail::Sendmail;
use strict;

my $BBSRPTEST = "bbsrptest.exe";
my $SMTP_SERVER = "smtp.server.com";
my $NO_OF_BES_SERVERS = 1;
my $EMAIL_TO = 'email@domain.com,anotheremail@domain.com';
my $EMAIL_FROM = 'email@domain.com';

my $output = `$BBSRPTEST`;

my $success_count = () = $output =~ /Successful/gs;
```

```
if ($success_count < $NO_OF_BES_SERVERS) {
    my $failures = $NO_OF_BES_SERVERS - $success_count;
    my %mail_options = (
        To       => $EMAIL_TO,
        From     => $EMAIL_FROM,
        Subject  => "[$ENV{COMPUTERNAME}] BB SRP Down!",
        Body     => "There was $failures SRP failures.  Output was:\n$output\n",
    );
    sendmail(%mail_options);
}
```

This Perl code uses the Mail::Sendmail Perl module, which doesn't come with the default distribution from ActiveState (which must be downloaded from *http://www.activestate.com/store/languages/register.plex?id=ActivePerl*). You'll need to install it by typing **ppm install Mail::Sendmail** from a command prompt. Type the following code into Notepad (or your favorite text editor) and save it as *bbsrptest.pl*. Of course, you will need to modify the variables at the top of the file.

Run the Code

To run the code, execute the following line from a command prompt in the directory where you saved the script.

```
C:\temp>perl bbsrptest.pl
```

The *bbsrptest* utility makes connection attempts for each BES you've configured on your server. This script counts the number of times the word "Successful" appears in the output of the *bbsrptest* command. If the count is less than the number of BES instances (which you've defined in the $NO_OF_BES_SERVERS variable), it will send an email to the address(es) you've specified in the $EMAIL_TO variable.

If you are a BlackBerry user and you've configured profiles on your device to receive alerts **[Hack #24]**, you'll probably want to add another email address in addition to email that delivers through your BES. In the event that the *bbsrptest* does fail and the script sends an alert, your device won't receive it if your BES is not operational. Consider adding the SMS email address for your device, such as *MOBILENUMBER*@mycingular.com.

Ping a User's BlackBerry

HACK
#76

Send a specially formatted email message that a BlackBerry will automatically confirm delivery of.

As a BlackBerry administrator, you often get calls from users claiming they are not receiving messages on their devices. Of course, there are several reasons a user might find himself in this situation; for example, he could be in an area that has no wireless coverage. Surprisingly, often a BlackBerry Enterprise Server user's device will be cradled with the "Disable email redirection while your handheld is connected" option selected in the user's Redirector Settings.

It would be nice if the BlackBerry operating system could confirm receipt of an email via an autoreply message. It turns out there is such a feature that comes in quite handy.

Send the Message

Just compose a message with the string <confirm> in the beginning of the subject, as shown in Figure 7-8. When the BlackBerry device receives a message with this in the beginning of the Subject field, it knows to send a reply to the sender with a confirmation that it has received the message (see Figure 7-9).

Figure 7-8. Sending a message that will be confirmed

Figure 7-9. Delivery to handheld confirmed

This feature is also good for sending important messages to users who are traveling and may be out of the coverage area. The reply confirms that they have indeed received the message to their handheld—of course, there are no guarantees that the user will read the message, however.

For BES users, you can request a read receipt that the BlackBerry will respect when the user reads the message.

Hack the Hack

There is another option you can include in the subject that controls the behavior of messages sent to the BlackBerry. You can add the removeondelivery option to the message subject to have the handheld delete the message immediately after receipt. This is useful to unobtrusively send a test message to a user without having the user actually see the message. Use the subject as follows (notice the dollar sign preceding the first option when multiple options are used):

```
<$confirm,removeondelivery> Actual Message Subject
```

> If you're using a version of the BlackBerry Enterprise Server earlier than 4.0.2, the message will be deleted on the server inbox but will remain on the handheld. This is because the BES performs a "hard delete" on the message, bypassing the Deleted Items folder entirely. On a pre-4.0.2 BES server, only messages that are "soft deleted" (that is, moved to the Deleted Items folder) are deleted from the handheld when Wireless Reconciliation is turned on. Starting with 4.0.2, the BES is supposed to sync the hard deletes as well as soft deletes.

HACK #77   Track Message Delivery Time

Use a built-in feature of the BlackBerry to monitor message delivery times.

How can you determine whether your BlackBerry platform is working and delivery times are "normal"? This is not an easy question to answer. You could send yourself a test message and manually watch your device for it to arrive. However, a successful test may indicate that the BlackBerry Enterprise Server that your device is homed on is functioning properly, but there may be other BES servers in your architecture that aren't functioning optimally.

Also, if there is a delay in receiving your test message, is the delay a normal duration or is it a significant delay that requires further troubleshooting?

You can use the built-in message delivery confirmation feature [Hack #76] to build a simple delivery time–monitoring tool to create a benchmark for message delivery times in your BlackBerry platform.

Requirements for This Hack

You'll need a functioning BlackBerry to which messages will be sent. This can be your own device, although it's best if you obtain another device just for this purpose. Ideally, the device could be stored in a secure place that has consistent wireless coverage—otherwise, your delivery times will be skewed as the device goes in and out of wireless coverage.

You will also need an automated way to send an email through some point in your network. It's best to send the test messages through the outermost point possible from your device—perhaps an external facing SMTP mail gateway that can deliver a message to your mail server. This allows your delivery times to reflect a worst-case scenario. On the other hand, you may want to confine your data as close as possible to your BlackBerry infrastructure—in effect removing as many factors as possible from the equation. In this case, you should relay the messages through the closest point to your mail server.

Send a Test Message

Here is some Perl code to send the message through an SMTP server. It uses the Mail::Sendmail module that doesn't come in the default Perl installation (despite the word Sendmail in the module, this module is platform independent and does not require Sendmail). You can install it using CPAN by typing perl -MCPAN -e "Mail::Sendmail" from a command prompt. Of course, you will need to plug in the values specific to your environment.

```perl
#! perl -w
use Mail::Sendmail;
use strict;

my $smtp_server = "mail.server.com"; # change to your SMTP server
my $mail_address = 'device@domain.com'; # should be device's email
my $from_address = 'monitor@domain.com'; # change to monitor mailbox
my $now = time( );

my %msg_options = (
    To => $mail_address,
    From => $from_address,
    Subject => "<\$confirm,removeondelivery> BB Delivery Time: $now",
    Body => "Message for delivery time monitoring",
);

sendmail(%msg_options);
```

Calculate the Delivery Time

Once the message is sent to the BlackBerry device, you'll receive the delivery confirmation email at the address specified in $from_address. The subject of the message is always "BlackBerry Delivery Confirmation," and the subject of your original sent message is in the message body. This allows you to check the confirmation message and compute the round-trip delivery time. Just take the received time of the confirmation, convert it to epoch time, and subtract the value of the epoch time that the original message was sent (I put it in the subject line as $now).

Here is the code I use to check the mailbox, perform the calculations on the delivery probe messages I sent using the first part of the script, and then delete the message. You'll need to change the values of $pop_server, $username, and $password to make them specific to your environment. Although it is possible to check the mailbox using a variety of protocols, I chose POP3 because of its broad support and ease of use.

```perl
#! perl -w
use strict;
use Time::Local;
use Mail::POP3Client;

my $pop3_server = "mail.server.com";
my $username = "username";
my $password = "password";

my $pop = Mail::POP3Client->new(
    USER        => $username,
    PASSWORD    => $password,
    HOST        => $pop3_server,
);

for (my $i = 1; $i <= $pop->Count(); $i++) {
    my $header = $pop->Head($i);
    if ($header =~ /Subject: BlackBerry Delivery Confirmation/) {
        my $body = $pop->Body($i);
        my $received_time = "";
        ($received_time) = ($header =~ /Received: from .*?; (.+?)\n/s);
        my $probe_sent = "";
        ($probe_sent) = ($body =~ /BB Delivery Time: (\d+)/);
        my $received_epoch = get_epoch($received_time);
        my $duration = $received_epoch - $probe_sent;
        print "Delivery took $duration seconds for" .
            "probe sent at $probe_sent.\n";
        $pop->Delete($i);
    }
}
$pop->Close;

sub get_time_stamp {
    my ($epoch) = @_;

    my ($sec,$min,$hour,$mday,$mon,$year,$wday,$yday,$isdst) =
        localtime($epoch);
    my $yyyymmdd = sprintf("%04d%02d%02d",$year+1900,$mon+1,$mday);
    my $time_stamp = sprintf ("%04d-%02d-%02d %02d:%02d:%02d",
        $year+1900,$mon+1,$mday,$hour,$min,$sec);
}

sub get_epoch {
    my ($smtp_timestamp) = @_;
    my @parts = split / /, $smtp_timestamp;
```

```
my $mday = $parts[1];
my $mon = $parts[2];
my $year = $parts[3];
my ($hour,$minute,$second) = split /:/, $parts[4];
my $tz = $parts[5];

my %month = (
    Jan   => 1,
    Feb   => 2,
    Mar   => 3,
    Apr   => 4,
    May   => 5,
    Jun   => 6,
    Jul   => 7,
    Aug   => 8,
    Sep   => 9,
    Oct   => 10,
    Nov   => 11,
    Dec   => 12,
);

my $epoch_wo_tz = timegm($second,$minute,$hour,$mday,
    $month{$mon}-1,$year);
$tz = $tz / 100;
my $epoch = $epoch_wo_tz - ($tz * 60 * 60);
}
```

Because the server where your mail resides stamps the received time on the message, you won't have to worry about when you run the delivery time calculation portion of the hack. All the data points you need to calculate the delivery time are in the message as it is received. You could even send out several test messages during the day, and then calculate them only once a day.

Hack the Hack

Once you've gathered enough data so that you are comfortable defining what a normal message delivery time is in your architecture, you could set a threshold that would warrant troubleshooting. Use that threshold and monitor the delivery times periodically, sending an alert if the threshold is exceeded. This type of proactive monitoring frees up your time for more important tasks. It's also nice to be automatically alerted of a potential problem so you can already have an idea of what the issue is by the time your users come calling.

Create Alerts for Important Users

#78   Start working on your CEO's BlackBerry problem before she realizes there is one.

It is amazing how important BlackBerry has become to some organizations. More and more executives now travel only with their BlackBerry, leaving their laptops behind. In many organizations, the BlackBerry has gone from nonexistent just a couple years ago to an absolutely mission-critical service today. With the high-profile users that are the typical users on a company's BlackBerry Enterprise Server, it is important that the administrators stay on their toes—it is amazing how quickly BlackBerry users will notice even a short delay in message delivery.

If there are certain important executives who rely heavily on their BlackBerrys, it is a good idea to set up alerting to detect when there are problems with delivery of messages for these users. Use the code in this hack to send alerts when certain users reach specific pending count thresholds.

The Code

```
use Text::CSV;
use Mail::Sendmail;
use strict;

my %USERS_TO_CHECK = (
    'email@domain.com' => -1, # pending count threshold for alerting
);

my $ALERTS_GO_TO = 'admin@domain.com';
my $ALERTS_FROM = 'admin@domain.com';
my $ALERTS_SUBJECT = "User Pending Count Threshold Exceeded";
my $SMTP_SERVER = "smtp.server.com";

my $BES_USER_ADMIN_CLIENT =
    "besuseradminclient.exe"; # change to your specific path
my %BES_SERVERS = (
    BES => "server.domain.com", # maps your BES name to the computer name
);
my $PASSWORD = "password"; # change to the password for BESUserAdmin
my $TEMP_DIR = "c:\\temp";

my $csv = Text::CSV->new;
foreach my $bes (sort keys %BES_SERVERS) {
    my $output_file = "$TEMP_DIR\\bes.$bes.users.csv";
    my $error_file = "$TEMP_DIR\\error.$bes.users.csv";
    my $system_command = "\"$BES_USER_ADMIN_CLIENT\" -n " .
      "$BES_SERVERS{$bes} -b $bes -p $PASSWORD -stats -users " .
      "> $output_file 2> $error_file";
    my $return_code = system $system_command;
    open OUTPUT, $output_file || die "Can't open output: $!\n";
```

```
my $count = 0;
while (<OUTPUT>) {
    chomp;
    $csv->parse($_);
    my @fields = $csv->fields;
    my $email = $fields[17];
    $email =~ s/^smtp://i;
    my $pending_count = $fields[9];
    if ($USERS_TO_CHECK{$email}) {
        # "important" user to check
        my $pending_threshold = $USERS_TO_CHECK{$email};
        if ($pending_count > $pending_threshold) {
            print "Sending alert for $email\n";
            my %message_options = (
                To      => $ALERTS_GO_TO,
                From    => $ALERTS_FROM,
                Subject   => $ALERTS_SUBJECT,
                Body    => "Alert threshold exceeded for $email " .
                        "($pending_count)\n",
                smtp    => $SMTP_SERVER,
            );
            sendmail(%message_options);
        } else {
            print "Pending count for $email was only $pending_count.\n";
        }
    }
}
close OUTPUT;
}
```

This Perl code uses the Mail::Sendmail and Text::CSV Perl modules, which don't come with the default distribution from ActiveState (which you'll need to have installed from *http://www.activestate.com/store/languages/register.plex?id=ActivePerl*). You'll need to install it by typing **ppm install Mail::Sendmail** and **ppm install Text::CSV** from a command prompt. Type the following code into Notepad (or your favorite text editor) and save it as *user_alert.pl*. Of course, you will need to modify the variables at the top of the file to customize it for your environment.

Run the Code

Run the code by bringing up a command prompt and changing directory to the folder where you stored the file. Type the following code to run the script:

```
C:\>perl user_alert.pl
```

Output

In the event that the pending count threshold for any of the users you specified is reached, the script will fire off an email to the address specified in the

$ALERTS_GO_TO variable. If the threshold is not met, you'll get console output similar to the following for each user you're checking:

```
Pending count for dmabe@runningland.com was only 0.
```

If you're testing, you can set the threshold to −1 so that the criteria will be met every time and the email will be fired off.

Move the Attachment Service

HACK
#79

Because the Attachment Service is so resource intensive, it's a good idea to run it on dedicated hardware.

The BlackBerry Attachment Service runs on each BlackBerry Enterprise Server you install. Most of the time it does a nice job of converting a variety of attachments into a format that is readable on your device. However, every so often you'll find a certain type of strange attachment sends the service to its knees. When this happens, it can impact the entire BlackBerry server, tying up resources and preventing other services from running optimally. In fact, there are instances when the Attachment Service has prevented any communications to or from handhelds while it chugs along consuming all the server's CPU time.

The BlackBerry Enterprise Server provides a way to point a BlackBerry Enterprise Server to another machine for attachment services. By moving the Attachment Service to another machine with either a smaller number of users or none at all, you isolate it from the rest of your BlackBerry platform. When a strangely formatted attachment is read that causes the resources to spike in the arrangement, your BlackBerry users won't even notice (unless, of course, they try to read an attachment while the remote Attachment Service is already pegged!).

The nice thing about moving the Attachment Service is that although a full BlackBerry Enterprise Server installation is required on the remote machine, the actual BlackBerry Service doesn't need to be running. Because of this, you won't need to purchase another SRP from Research In Motion to run the Attachment Service on dedicated hardware.

Set Up the Remote Attachment Service

On the remote machine, install the BlackBerry Enterprise Server as you normally would. When you are finished, simply set all the BlackBerry services except for the BlackBerry Attachment Service to Disabled so they won't start automatically. Figure 7-10 shows setting the service to the Disabled state. Repeat this process for each of the BlackBerry services except for the Black-Berry Attachment Service.

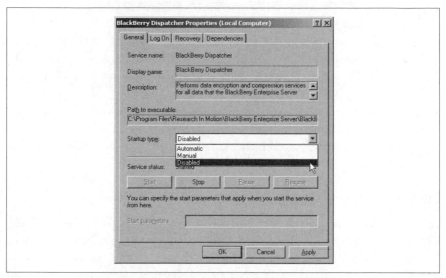

Figure 7-10. Disabling BlackBerry services on the remote machine

Point to the Remote Service in 3.6

On each machine with users on it, you'll need to tell the local BlackBerry Enterprise Server where to look for attachment processing. In BlackBerry Enterprise Server 3.6 and earlier, there is a tool for configuring the Attachment Service on the BES server. Go to Start → Program Files → BlackBerry Enterprise Server → Attachment Configuration Tool. Enter the new attachment connector settings for your environment as shown in Figure 7-11.

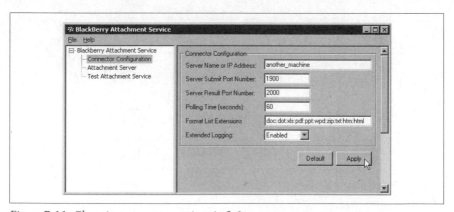

Figure 7-11. Changing connector settings in 3.6

After you click Apply to save your changes to the configuration, you are prompted to stop and restart the server, as shown in Figure 7-12.

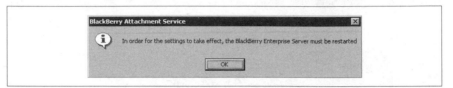

Figure 7-12. Restart the BlackBerry Server service in 3.6 for your changes to take effect

Although this warning gives the impression that the entire computer must be restarted, in fact, only the BlackBerry Server service actually needs to be restarted for the settings to take effect. You'll need to do this manually after clicking OK.

Point to the Remote Service in 4.0

In BlackBerry Enterprise Server 4.0, go to Start → Program Files → Black-Berry Enterprise Server → BlackBerry Server Configuration. Click the Attachment Server tab and make sure the Connector Configuration option is selected. By default, the server is set to localhost. Change the Server field under the Connector Configuration section to the hostname of the computer to which you'd like this BES to point for attachment processing, as shown in Figure 7-13.

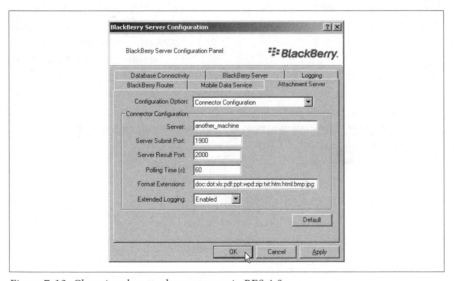

Figure 7-13. Changing the attachment server in BES 4.0

Although changing the settings in 4.0 does not prompt you to restart services as it does in 3.6, you still have to restart the BlackBerry Dispatcher service for these changes to take effect.

HACK #80 Monitor Attachment Usage

Determine how heavily your users are accessing the Attachment Service.

Attachment reading with the BlackBerry is a very useful feature of the Black-Berry Enterprise Server. Having the ability to read the contents of attachments comes in very handy when you don't have access to a desktop computer with office productivity software installed.

This power comes at a price—the BlackBerry Attachment Service is a resource-intensive beast that requires a sizeable amount of system resources to work effectively. As a BlackBerry administrator, it is a good idea to keep track of Attachment Service usage over time. As more and more of your users discover the functionality, your hardware requirements might change and you may want to upgrade the server or move the service to its own dedicated hardware [Hack #79].

Turn On Attachment Service Logging

Attachment logging is disabled by default in both BlackBerry Enterprise Server 3.6 and 4.0. You can enable it by logging onto your BlackBerry Enterprise Server and going to Start → Programs → BlackBerry Enterprise Server → Attachment Service Configuration Tool. Click on the Connector Configuration node on the left size of the dialog box. As shown in Figure 7-14, enable the Extended Logging feature and click Apply. After setting to Enabled, you'll be prompted with a dialog box that indicates you will need to stop and start the BlackBerry Service for the setting to take effect.

In Version 4.0, use the BlackBerry Server Configuration Tool to set the Debug Log Level to 5 for the BlackBerry Attachment Service and BlackBerry Attachment Conversion. Due to a bug in the current version of 4.0, you'll also need to modify the registry to enable the level of logging required to see the attachment names in the debug logs. Set `HKLM\Software\Research In Motion\BBAttachServer\BBAttachBESExtension\BBAttachEnableExtensionLog` to 1, set `HKLM\Software\Research In Motion\BBAttachServer\BBAttachBESExtension\BBAttachServerLogLevel` to 5, and then set `HKLM\System\CurrentControlSet\Services\BBAttachServer\Parameters\EnableLog` to 5.

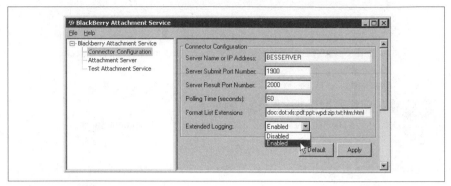

Figure 7-14. Enabling Extended Logging in 3.6

Debug Log Entries

Attachment Service usage gets logged to the standard debug log on the BlackBerry Enterprise Server that the reader's device is homed on. A typical set of log entries for BES 3.6 that occur when someone reads an attachment is as follows:

```
[40000] (04/13 10:56:10):{0x3DC} TransID=1241871075, AttachmentTempFileName=C:\
DOCUME~1\bessvc2\LOCALS~1\Temp\bes38, Attachment data successfully retrieved,
Success=0x0000000C
[40000] (04/13 10:56:10):{0x3DC} TransID=1241871075, Entering the _
ConvertAttachment method attachment=sales presentation.ppt, Success=0x00000002
[40000] (04/13 10:56:10):{0x3DC} TransID=1241871075, (dmabe@runningland.com)
Operation completed successfully - Attachment=sales presentation.ppt,
Success=0x00000000
[40000] (04/13 10:56:10):{0x3DC} TransID=1241871075, Leaving the _
ConvertAttachment method attachment=sales presentation.ppt, Success=0x00000004
[40000] (04/13 10:56:10):{0x3DC} TransID=1241871075, Atomic attachment
conversion request completed successfully attachment=sales presentation.ppt,
Success=0x00000005
[40000] (04/13 10:56:10):{0x3DC} TransID=1241871075, Leaving the
ExtendedMoreRequest2 method, Success=0x00000001
[30328] (04/13 10:56:10):{0x3DC} {dmabe@runningland.com} The extension
returned MORE_COMPLETE, for the Extended More Request, for transid
1241871075.
[40439] (04/13 10:56:10):{0x3DC} {dmabe@runningland.com} Actual MORE data
available, size=2783, Tag=2113806225, TransactionId=1241871075
[40298] (04/13 10:56:10):{0x3DC} {dmabe@runningland.com} Sending MORE to
device, RefId=-712861129, PartId=0, Offset=0, Length=3000
[40140] (04/13 10:56:10):{0x3DC} Encrypting using key ID 0B2746A34+G+i
[40583] (04/13 10:56:10):{0x3DC} {dmabe@runningland.com} Sending message to
device, Size=1503, Tag=366484, TransactionId=-1065991950
[40279] (04/13 10:56:10):{0x3DC} {dmabe@runningland.com} SubmitToRelaySendQ,
Tag=366484
[40279] (04/13 10:56:10):{0x3DC} {dmabe@runningland.com} SubmitToRelaySendQ,
Tag=2113806225
```

```
[40000] (04/13 10:56:10):{0xAC0} [SRP] Send data, Tag=366484
[40000] (04/13 10:56:10):{0xAC0} [SRP] Send status DATA_ACCEPTED,
Tag=2113806225
```

> BlackBerry Enterprise Server Version 4.0 and greater has a
> different format for the attachment viewing in its debug log.
> You'll find the entries in the *_MAGT_* logfiles. Despite the
> difference in formatting, the code in this hack will work for a
> 4.0 log or a 3.6 log.

The previous excerpt shows that a user (*dmabe@runningland.com*) performed
an attachment read on a Microsoft PowerPoint file named *sales presentation.ppt*.
You can use a Perl script to go through the file looking for lines similar to this:

```
[40000] (04/13 10:56:10):{0x3DC} TransID=1241871075, (dmabe@runningland.com)
Operation completed successfully - Attachment=sales presentation.ppt,
Success=0x00000000
```

When you read an attachment from a BlackBerry, you don't retrieve the entire
contents of it all at once. Similar to reading a long email message, your device
retrieves the first section (actually the first few bytes) of the attachment and
places it in a buffer. When you approach the end of the buffer, your device
then retrieves the next section of the attachment and puts it into the buffer.
This process happens fairly quickly, so it's possible you may not even realize
the whole attachment isn't retrieved at once at the beginning.

This also means that every time a device retrieves "more" of the attachment,
another line gets logged to the file similar to the previous line. You'll need to
account for this in the code so that you don't count these "more retrievals"
as additional attachment reads.

The Code

```
use strict;

my $file = "path to debug log";

my $attachments = {};
my %refids = ();

open FILE, $file;
while (<FILE>) {
   chomp;
   my ($idtext,undef) = split / /;
   my $id = substr $idtext, 1, 5;
   if ($id == 40182) {
      my ($refid) = (/{(0x[^}]{3,4})}/);
      my ($email) = (/} {([^@]+@[^}]+)}/);
      $refids{$refid} = $email;
```

```
    }
    if ($id == 40000 and
        / Operation completed successfully - Attachment=/) {
            # 3.6 log
        my ($email) = (/[{\(]([^\({\@]+\@[^}\)]+)[}\)]/);
        my ($filename) = (/\)\sOperation\scompleted
            \ssuccessfully\s-\sAttachment=(.+),\sSuccess=/x);
        my $timestamp = substr $_, 9, 14;

        if ($attachments->{$email}->{$filename}) {
            $attachments->{$email}->{$filename}->{endtime} = $timestamp;
        } else {
            $attachments->{$email}->{$filename}->{starttime} = $timestamp;
        }
    }
    if ($id == 30000 and /, Attachment=([^,]+), Attachment conversion/) {
        # 4.0 log
        my $filename = $1;
        my ($refid) = (/{(0x[^}]{3,4})}/);
        my $email = $refids{$refid};
        my $timestamp = substr $_, 9, 14;

        if ($attachments->{$email}->{$filename}) {
            $attachments->{$email}->{$filename}->{endtime} = $timestamp;
        } else {
            $attachments->{$email}->{$filename}->{starttime} = $timestamp;
        }
    }
}

foreach my $email (keys %{ $attachments }) {
    foreach my $attachment (keys %{ $attachments->{$email} }) {
        my $start_time = $$attachments{$email}{$attachment}{starttime};
        my $end_time = $$attachments{$email}{$attachment}{endtime};
        print "$email\t$attachment\t$start_time\t$end_time\n";
    }
}
```

Run the Code

Save the previous code in a text file called *bbattach.pl* and run the following command from a command prompt.

```
C:>perl bbattach.pl
```

Running the code will produce output similar to the following:

```
dmabe@runningland.com sales presentation.ppt 04/13 10:56:10 04/13 10:59:10
```

It parses your debug log and tracks the first and last read times of attachments read by your users. In the previous example, user *dmabe@runningland.com* started reading an attachment called *sales presentation.ppt* at 10:56 A.M., and the last retrieval of data occurred at 10:59 A.M.

Export User Stats to a Text File

Use the BlackBerry User Administration client to export the current users on your server to a text file.

Research In Motion supplies a far underused utility for performing user administration from a command prompt. The BlackBerry User Administration client can be used to add users to the BES, remove users, and export the users along with some BlackBerry specific statistics to a text file suitable for viewing and manipulating in a spreadsheet program. Starting with BES 4.0, BlackBerry User Administration has been enhanced to perform several additional tasks that were not previously possible from the command line.

The BlackBerry User Administration software used to come as an optional component of the BES installation. Starting with BES 4.0, it is no longer included with the server software, but is still available for download as part of the BlackBerry Resource Kit (*http://www.blackberry.com/support/downloads/resourcekit.shtml*). There is a service that must be installed on your BlackBerry Enterprise Server and a client utility that is also installed on the server, but can be copied to other servers without BES installed.

Install the Service

For versions of BES 3.6 and earlier, you can install the BlackBerry User Administration client from the BlackBerry Enterprise Server installation. You will need to create MAPI profiles on your machine similar to the profiles that you created for the BlackBerry Service account mailbox, or you can use the same profile as the service account. There is documentation that comes with the installation files.

For Versions 4.0 and greater, the BlackBerry User Administration client is no longer bundled with the server software, but with the BlackBerry Resource Kit. Simply unzip the archive to whatever location you'd like on the BlackBerry Enterprise Server, although a convenient directory in which to install it is alongside the BES install in:

C:\Program Files\Research In Motion\BlackBerry Enterprise Server\BRK

Export User Stats with the 4.0 Version

To export user statistics from a BES running 4.0 or greater, bring up a command prompt, *cd* to the directory where *besuseradminclient.exe* is located, and type the following command, substituting the values for your environment for the variables in all caps.

```
C:\>besuseradminclient -n COMPUTER -b BES -p PASSWORD -stats -users
```

At first glance, the output from the command doesn't look very helpful. You should get output similar to the following.

```
BlackBerry User Administration Client Version 4.0.1.8
Copyright (c) Research In Motion, Ltd. 1997-2004. All rights reserved.
Modification date: May  2 2005

besuseradminclient -n COMPUTER -b BES -p PASSWORD -users -stats

BESUserAdmin::main - Log Start
BESUserAdmin::DoListUsersStats - Listing users stats...
User name,MailBoxDN,ServerDN,PIN,Device Type,State,Message
Server,Forwarded,Sent,Pending,Filtered,Expired,Status,Last fwd time,Last sent
time,Last contact time (h),Last result,SMTP Address,Mobile Data Service,OTA
Calendar,ITPolicy name,ITPolicy status,ITPolicy time sent,ITPolicy time
received,Wireless Message Reconciliation
"Mabe, David M",/o=org/ou=site/cn=Recipients/cn=dmabe,/o=org/ou=site/
cn=Configuration/cn=Servers/cn=WIN2K3/cn=Microsoft Private
MDB,20000008,GPRS,Enabled,WIN2K3,14,0,0,0,0,(**WIN2K3) Running,Wed Jun 01 13:14:
52 2005,Wed Jun 01 13:14:35 2005,2.77,,SMTP:dmabe@domain.
com,Enabled,Enabled,Default,Applied successfully,Mon May 30 10:13:32 2005,Mon
May 30 10:13:33 2005,Enabled
BESUserAdmin::DoListUsersStats - ...done.
BESUserAdmin::main - Log End
```

The command prints its log messages to standard error while printing the actual user statistics to standard out. This allows you to separate the different outputs fairly easily. The following command redirects standard error (the debug messages) to the Windows equivalent of Unix's */dev/null* (i.e., the bit bucket) and redirects the actual output we want to a local file (you must type this command all on one line):

```
besuseradminclient -n COMPUTER -b BES -p PASSWORD -stats
-users > users.csv 2> NUL
```

This command should print no output to the screen and it should have created a file called *users.csv* in the current working directory. This file should contain the user statistics from all users on the BES. You can repeat this command for each BES in your environment if you have multiple BES instances.

Export User Stats with the 3.6 Version

The command to run to export user statistics with the 3.6 (and earlier) version of BES is significantly different. The filename is different along with the output format and command-line options that are required. To create an export, run the following command in a prompt in the same directory as the *bbuseradminclient.exe* file (note the different filename in this version).

```
C:\>bbuseradminclient -n COMPUTER -b BES -x
```

You should get output similar to the following:

```
BlackBerry User Administration Client, Version 2.1
Copyright (c) Research In Motion, Ltd. 1997-2002. All rights reserved.
Modification date: Oct 19 2004
Users currently on BES BlackBerry Server:

Mabe, David M
    PIN: 20000008
    Device Type: GPRS
    State: Enabled
    Exchange Server: EXCHANGE_SERVER
    Forwarded: 8042
    Sent: 77
    Pending: 0
    Filtered: 59
    Expired: 0
    Status: Running
    Last fwd time: Wed Jun 01 07:02:32 2005
    Last sent time: Wed Jun 01 06:47:02 2005
    Last contact time: 00:03:39
    Last result: Delivered to handheld
    External Services: Enabled
    OTA Calendar: Enabled
    ITPolicy name: Default
    ITPolicy status: Applied successfully
    ITPolicy time sent: Fri Feb 18 22:22:27 2005
    ITPolicy time received: Tue May 24 10:40:04 2005
    Wireless Email Reconciliation: Enabled

bbuseradminclient.exe finished
```

It will list the users on the BES you've specified and print out a bunch of statistics that pertain to the BlackBerry service. At first glance, this looks like a pretty readable and convenient format. However, the instant you try to bring up the output in a spreadsheet program to do something useful with it, you'll realize what a silly format this is.

Parse the Output

If it can't be automated, I'm not interested in it. That's why I wrote some Perl code to convert this ridiculous format into a more reasonable tab-delimited format. Type the following text into your favorite text editor and save the file as *bbuseradmin.pl*. You'll probably need to substitute the correct path for the $BB_USER_ADMIN variable unless it is in your path.

```
use Getopt::Long;
use strict;

# change this if it's not in your path
my $BB_USER_ADMIN = "bbuseradminclient.exe";
```

```perl
my $BES = "";
my $COMPUTER = "";
my $FILE = "";
my $INPUT = "";
GetOptions(
    "bes=s"         => \$BES,
    "computer=s"    => \$COMPUTER,
    "file=s"        => \$FILE,
    "input=s"       => \$INPUT,
);

my $output_file = $FILE || "output.txt";
my $temp_file = $INPUT || "temp.txt";
my $command = "\"$BB_USER_ADMIN\" -n $COMPUTER -s $BES -u > $temp_file";

if (not $INPUT) {
    my $return_code = system $command;

    if ($return_code) {
        print "Uh-oh.  The command was unsuccessful.  " .
            "Return code was: $return_code\n";
        die;
    } else {
        print "Command was successful.  Creating csv file...\n";
    }
}

my @header_row = (
    "Display Name",
    "PIN",
    "Type",
    "State",
    "Server",
    "Forwarded",
    "Sent",
    "Pending",
    "Filtered",
    "Expired",
    "Status",
    "Last fwd time",
    "Last sent time",
    "Last contact time",
    "Last result",
    "OTA Calendar",
    "External Services",
    "ITPolicy name",
    "ITPolicy status",
    "ITPolicy time sent",
    "ITPolicy time received",
    "Wireless Email Reconciliation",
);

if (open FILE, $temp_file) {
```

```
    my $line_count = 0;
    open OUTPUT,">$output_file" ||
        die "Can't open $output_file for write: $!";
    print OUTPUT join("\t", @header_row),"\n";
    my $cached_text = '';
    my $names = 0;
    while (<FILE>) {
        chomp;
        $line_count++;
        next if ($line_count < 6);
        if (/bbuseradminclient\.exe finished/) {
            last;
        }

        if (not $_) {
            $cached_text =~ s/\t$//;
            next;
        }

        if (/^[^\t]/) {
            $names++;
            $cached_text .= "\n" if $names != 1;
            $cached_text .= $_ . "\t";
        } else {
            $_ =~ s/^\t[^:]+:\s?//;
            $cached_text .= $_ . "\t";
        }
    }
    print OUTPUT $cached_text;
    close OUTPUT;
    close FILE;
}
```

To run the code, type the following command, substituting the appropriate command-line options for your environment.

```
C:\>perl bbuseradmin.pl --bes BES --computer COMPUTER
```

This command runs the *bbuseradminclient.exe* utility with the appropriate options and redirects output to a temporary file. It then opens the temporary file, parses the text for each user, and massages it into a tab-delimited line. It writes the output to a file named *output.txt* in the current directory. You can easily open the resulting file in Microsoft Excel.

HACK #82 Create a Web Interface for User Administration

Using the BES User Admin service, create a web interface for adding and removing users.

Often it's not so convenient to have to use the BlackBerry Manager to add, remove, or just gather statistics about your users. Until very recently (Version BES 4.0 SP1 Hot Fix 2, in fact), the BlackBerry Manager GUI took painfully long to load the user list, because it has to make a connection to each user's mailbox to retrieve statistics from the hidden folders where they are stored. This made remote administration across even a fairly sizable network link a real chore.

With the BES User Admin service, you can set up a web interface through which to add and remove users and view statistics from a web browser. Not only can you manage BlackBerry users remotely without installing the BlackBerry Manager MMC, but it is quicker than using the BlackBerry Manager remotely because the BES is retrieving the statistics and then just returning them to your web browser over HTTP.

Set Up the Web Server

Because it requires that a web server be installed, your BES is probably not the best place to run this hack. With the critical nature of your BES, you should try to minimize any extraneous services on that machine. It is best to copy the BES User Admin client executable and DLLs to another machine and install a web server there.

To install the web interface, you'll need to install ActiveState Perl on your Windows web server if it's not there already. I used Perl Version 5.8.6, available at the ActiveState web site (*http://www.activestate.com/store/languages/ register.plex?id=ActivePerl*), although any recent version should work. I used IIS, but the Apache web server should work just fine. There are some other steps to take to ensure this works properly.

- Make sure *.pl* files are mapped to Perl in your web server. If you install ActivePerl after you install IIS, the ActivePerl installer can configure this for you.

- Use a combination of NTFS permissions on the directory and NTLM authentication on the web site to restrict access to the web site.

- Make sure the temp directory you use is created and writeable by anyone that will be using the web site.

- Install the Text::CSV Perl module on the web server using the following command: ppm install Text::CSV.

The Code

```perl
#!perl -w

use strict;
use CGI;
use Text::CSV;

my $BES_USER_ADMIN_CLIENT =
   "besuseradminclient.exe"; # change to your specific path
```

```perl
my %BES_SERVERS = (
    BES_NAME => "COMPUTER_NAME", # this maps your BES name
                               # to the computer name
);
my $PASSWORD = "password"; # change to the password for BESUserAdmin
my $TEMP_DIR = "c:\\temp";

my $cgi = CGI->new;
print $cgi->header;
print $cgi->start_html("BES User Administration");

print <<"END";
<style type="text/css">
body {
  font-size: 9pt;
  font-family: arial;
}
</style>
END

print $cgi->h1("BES User Administration");

my $op = $cgi->param("op");

if ($op eq 'remove') {
    my $bes = $cgi->param("bes");
    my $user = $cgi->param("user");
    my $error_file = "$TEMP_DIR\\error.$bes.remove.csv";
    my $system_command = "\"$BES_USER_ADMIN_CLIENT\" " .
        "-n $BES_SERVERS{$bes} " .
        "-b $bes -p $PASSWORD -delete -u $user 2> $error_file";
    if (system $system_command == 0) {
        print "<p>Successfully removed $user from $bes</p>\n";
    } else {
        print "<p>ERROR removing $user from $bes.</p>\n";
    }
} elsif ($op eq 'add') {
    my $bes = $cgi->param("bes");
    my $user = $cgi->param("user");
    my $activation_password = $cgi->param("act_password");
    my $error_file = "$TEMP_DIR\\error.$bes.add.csv";
    my $system_command = "\"$BES_USER_ADMIN_CLIENT\" -n $BES_SERVERS{$bes} " .
        "-b $bes -p $PASSWORD -add -u $user " .
        "-w $activation_password 2> $error_file";
    if (system $system_command == 0) {
        print "<p>Successfully added $user from $bes</p>\n";
    } else {
        print "<p>ERROR adding $user from $bes.</p>\n";
    }
}

my $csv = Text::CSV->new;
foreach my $bes (sort keys %BES_SERVERS) {
```

```perl
    my $output_file = "$TEMP_DIR\\bes.$bes.users.csv";
    my $error_file = "$TEMP_DIR\\error.$bes.users.csv";
    my $system_command = "\"$BES_USER_ADMIN_CLIENT\" " .
      "-n $BES_SERVERS{$bes} " .
      "-b $bes -p $PASSWORD -stats -users > $output_file 2> $error_file";
    my $return_code = system $system_command;
    open OUTPUT, $output_file || die "Can't open output: $!\n";
    print "<table border=\"1\">\n";
    my $count = 0;
    while (<OUTPUT>) {
        chomp;
        $csv->parse($_);
        my @fields = $csv->fields;
        splice @fields, 1, 2; # remove the really long fields
        if ($count == 0) {
            print
                "<tr><td><strong>Remove</strong></td><td><strong>",
                join("</strong></td><td><strong>",@fields),
                "</strong></td></tr>\n";
        } else {
            print
                "<tr><td><a href=\"./besadmin.pl?op=remove" .
                "&amp;user=$fields[0]&amp;bes=$bes\">Remove</a></td><td>",
                join("</td><td>",@fields),
                "</td></tr>\n";
        }
        $count++;
    }
    print "</table>\n";
    close OUTPUT;
}

print <<"END";
<p>
<form action="./besadmin.pl" method="post">
<input type="text" name="user" value="Username" />
<input type="text" name="act_password"
value="Activation Password" />
<input type="hidden" name="op" value="add" />
<select name="bes">
END

foreach my $bes (sort keys %BES_SERVERS) {
    print "\t<option value=\"$bes\">$bes</option>\n";
}

print <<"END";
</select>
<input type="submit" value="Add" value="Add" />
</form>
</p>
END

print $cgi->end_html;
```

Run the Code

Type the code into Notepad, your favorite editor, or download it from the book's web site (see the Preface for more information on getting and using the code in this book). The file has to be named *besadmin.pl* and must be placed in a virtual directory that can execute scripts.

Be sure to modify the variables at the top of the script to values that are specific to your environment. Once you've copied the file to your web server, try accessing it from a web browser using something similar to the following URL:

http://www.your.server.com/directory/besadmin.pl

Of course, your URL will be different. You should get a page that looks similar to Figure 7-15. Note that I've truncated many of the fields on the right side of the table.

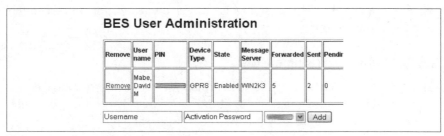

Figure 7-15. The BES Admin web interface

This site allows you to add and remove users to and from your BES servers, as well as view user-specific statistics. You can use the form at the bottom to add users and assign an activation password, which will allow them to be provisioned over the air.

Hack the Hack

With a little modification and some more coding, you could create a site that allows your users to enable themselves without any intervention from administrators. I've set up a similar site that handles all user account creations and ensures that users have obtained management approval before they are added to the BES. If you have multiple BES servers and sites, you could introduce some logic to your web site that determines the optimal BES to add users to, given the location of their mail servers.

Set Owner Info for All Devices

Set owner information on devices over the air with no user intervention.

When a password-protected device goes into the locked state after inactivity, it can display a custom message that identifies the owner of the device. This can be crucial if you have any devices that get lost—if there is no owner information specified in the device, there is no way for someone to know who owns the device to return it. With older versions of BlackBerry Enterprise Server, you had to rely on the user to input owner information in the Owner section of the Options program on his device. With BES 3.5, you could send the owner information over the air, but you had to use the BlackBerry Management tool's GUI to set the information for each user one at a time. If you are using BlackBerry Enterprise Server Version 4.0 or greater, you can use the BES User Admin tool to set owner information remotely from a command prompt.

Set the Owner Info Using the GUI

To set the owner information for a user using the BlackBerry Management Console, right-click on a user, go to IT Admin → Set Owner Info, and enter the name of the user and any extra information, as shown in Figure 7-16.

Figure 7-16. Setting owner information using the GUI

After the device is in a coverage area for a few seconds, it will pick up the change without any indication to the user. Figure 7-17 shows the device after retrieving the owner information from the server. This provides a good way for the owner to give a little bit of information to help recover a device in the event it is lost, without compromising any potentially proprietary information to the finder.

Figure 7-17. The owner information is displayed when the device is password locked

Set the Owner Info Using the Command Line

You can set owner information using the GUI, but your mouse hand is likely to lock up with rigor mortis if you have to do it for more than a few users. Once you've set up the BES User Admin service [Hack #81], you can use the following command to set the owner information from a command prompt. Of course, you'll have to substitute the information specific to your environment (be sure to type this all on one line).

```
besuseradminclient -n server_name -b bes_name -p password
-set_owner_info -u "/o=org/ou=site/cn=container/cn=alias"
-name "David M. Mabe" -info "If found call 919 960 5555"
```

The results of the command should be similar to the following:

```
BlackBerry User Administration Client Version 4.0.1.8
Copyright (c) Research In Motion, Ltd. 1997-2004. All rights reserved.
Modification date: May  2 2005

besuseradminclient -n server_name -b bes_name -p password
 -set_owner_info -u "/o=org/ou=site/cn=container/cn=alias"
 -name "David M. Mabe" -info "If found call 919 960 5555"

BESUserAdmin::main - Log Start
BESUserAdmin::DoSetOwnerInfo - ITAdmin SET_OWNER_INFO...
BESUserAdmin::DoSetOwnerInfo - ...done.
BESUserAdmin::main - Log End
```

Hack the Hack

While this is a bit easier on your mouse, you still have to complete the command one user at a time. Here is some Perl code to read from a tab-delimited input file and run the command once per line in the file. Set up your input file (*input.txt*) as follows:

```
Server_name   bes_name   account_info   owner_name   owner_info
```

The Perl code follows. You'll need to substitute your BES User Admin service password for the $PASSWORD variable. You may have to change the value of $BES_USER_ADMIN_CLIENT if it's in another directory and not in your path:

```
#! perl -w
use strict;
my $BES_USER_ADMIN_CLIENT = "besuseradminclient.exe";
my $PASSWORD = "password";
my $INPUT_FILE = "input.txt";

open INPUT, $INPUT_FILE || die "Can't find $INPUT_FILE: $!\n";
while (<INPUT>) {
    chomp;
    my ($server,$bes,$account,$owner,$info) = split /\t/;
    my $system_command = "\"$BES_USER_ADMIN_CLIENT\" -n $server " .
      "-b $bes " .
      "-p $PASSWORD -u $account -name \"$owner\" -info \"$info\"";
    my $return_code = system $system_command;
    if ($return_code) {
        print "There was an error for $account\n";
    } else {
        print "Success for $account.\n";
    }
}
close INPUT;
```

Make sure that *input.txt* is in the current directory, save the previous code in a file called *set_owner.pl*, and run the program by running this command.

```
C:>perl set_owner.pl
```

HACK #84 Query the BES 4.0 Database

Version 4.0 of the BES and the handheld added some excellent capabilities for reporting. Run reports based on model number, code version, and even applications installed on devices.

If you are the curious type and you ever looked into the schema of the Black-Berry Enterprise Server's database before 4.0, you probably noticed something striking. The database was hardly used for anything! Before IT policies were introduced as a feature, there were only two tables in the BESMgmt database. This stands in stark contrast to what exists in the 4.0 database schema. The 4.0 database is used to a much greater degree—my installation contains no fewer than 63 tables!

Where was all this stuff stored in previous versions? In many cases, the feature that the data applies to just didn't exist. Other types of data were stored in the BlackBerry service account mailbox in the hidden folders. With this shift to storing all data in SQL, this opens up all kinds of opportunities for running reports using this central database.

One nifty feature of 4.0 is that along with syncing all the different PIM items (mail messages, contacts, tasks, etc.) over the air, it also syncs other valuable information that administrators might find interesting. For example, the hand-held system software version number and the device model number are both stored in the database. Here's a script to produce a report on some of the more interesting statistics that BlackBerry administrators didn't have access to in previous versions. This requires that you be logged onto a computer with an account that has access to the BESMgmt database on the SQL server.

The Code

```perl
use DBI;
use strict;

my $SERVER = "bes.server.com";

my $DSN = "DBI:ODBC:Driver={SQL Server};Server=$SERVER;Database=BESMgmt";

my $dbh = DBI->connect($DSN);

if (not $dbh) {
    print $DBI::errstr,"\n";
    die;
}

my $sql =<<"END";
SELECT u.displayName,u.PIN,s.modelname,s.PlatformVer,s.AppsVer,
s.PhoneNumber,s.IMEI,s.HomeNetwork,s.PasswordEnabled,s.FlashSize
,s.ITPolicyName
FROM SyncDeviceMgmtSummary AS s INNER JOIN userconfig AS u
ON u.id = s.UserConfigId
ORDER BY u.displayName

END

my $st = $dbh->prepare($sql) || die $DBI::errstr;
$st->execute;

my @headings = qw(
    DisplayName
    PIN
    ModelName
    PlatformVer
    AppsVer
    PhoneNumber
    IMEI
    HomeNetwork
    PasswordEnabled
    FlashSize
    ITPolicyName
);

print join("\t", @headings),"\n";
while (my @fields = $st->fetchrow) {
    print join("\t", @fields),"\n";
}
```

The code has a few prerequisites. You have to install ActiveState Perl (*http://www.activestate.com*) and some optional Perl modules: DBI and DBD::ODBC. It's best to use the Perl Package Manager to install the required modules by running the following commands: `ppm install DBI` and `ppm install DBD::ODBC`.

Run the Code

Save the code in a file called *deviceinfo.pl*. From a command prompt, change directory (*cd*) to the directory where you saved the file and type the following command to run the code.

```
C:\>perl deviceinfo.pl
```

Output

In your command window, you should see a tab-delimited list of your BlackBerry users along with the following statistics for each.

- Display name of the user
- PIN
- Model number
- Platform version
- Application versions
- Device phone number
- Home network
- Password featured enabled
- Flash memory capacity
- Current BlackBerry IT policy in effect on device

To redirect the output of the script to a file named *output.txt*, simply run the following command:

```
C:\>perl deviceusers.pl > output.txt
```

This creates a tab-delimited file that can be easily viewed in a spreadsheet program such as Microsoft Excel.

The Web and MDS

Hacks 85–93

One of the most significant innovations in RIM's short history is the addition of the BlackBerry Browser to the operating system. In combination with the BlackBerry Enterprise Server's Mobile Data Service (MDS), instantly unlocked entire intranets along with mountains of rich corporate data made them available to the mobile user. As BlackBerry users catch on and start using the BlackBerry Browser, they'll quickly find room for improvement in your intranet.

Corporate networks are chock-full of lazily coded FrontPage web sites that were designed for ancient versions of Internet Explorer viewed with large monitors. How could those web developers have known their sites would one day be viewed on a mobile device from anywhere? To make those sites viewable on the BlackBerry, some sites will require minor tweaks [Hack #88]. For others, it might be easier to start from scratch. For especially time-sensitive data, you can push it [Hack #90] to your users' BlackBerry devices.

HACK #85 Control Access to Certain Sites

If your company uses a proxy server, don't tell MDS about it.

When the Mobile Data Service was first released with BlackBerry Enterprise Server Version 3.5, it had poor support for proxy servers. You couldn't configure it with an auto-proxy, which a lot of companies use to provide Internet access to users. At that time, you could only configure the server to either use a proxy for all HTTP requests or use a proxy for none. There wasn't any "no proxy" list that told the service to contact certain servers directly, bypassing the proxy. This was initially a show-stopper for many MDS deployments.

BES 3.6 solved some of these issues by allowing access to an auto-proxy, and MDS deployments were more plentiful; however, without any bypass proxy list some deployments were left choosing between allowing access to intranet or Internet sites but not both. With BES 4.0, there are significant enhancements to the level of control you have over specific URLs. If you are the security-conscious type and don't like your users having any fun at all, there are several different ways you can control what your users have access to with the Mobile Data Service.

Hack the Proxy Settings in BES 3.6

While the possibilities are limited with BES 3.6 compared to the feature set of 4.0, there are some tricks you can use to control access to URLs. If your company requires the use of a proxy server to access Internet sites, you could use that requirement to exclude any access to Internet sites from your BlackBerry users by simply not including the proxy configuration in your instance of MDS. This would instruct the Mobile Data Service to make a direct connection to all sites—the Internet site would be inaccessible (since the firewall denies any direct connection), while intranet sites could easily be reached. Figure 8-1 shows the settings page for the proxy configuration on a BES 3.6 server.

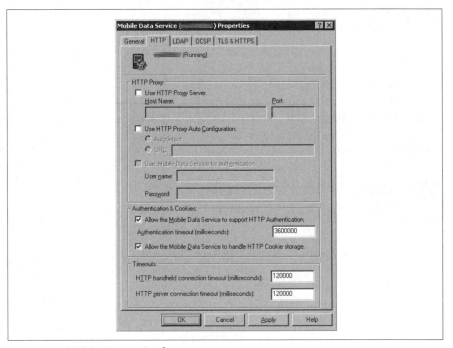

Figure 8-1. BES 3.6 Proxy Configuration settings

If your company uses a proxy that is configured to make HTTP requests only to Internet sites, you're not quite out of luck if you want your users to have access to both intranet and Internet sites. You could code a very simple proxy that knows which URLs define your intranet and could choose whether to use your "real" proxy to access intranet sites or contact intranet sites directly. Although this sounds like a complicated program, it can be accomplished with just a few lines of Perl. (Of course, what can't be?)

Use the Advanced Features in BES 4.0

The Mobile Data Service included with BlackBerry Enterprise Server 4.0 has features that the 3.6 version could only dream of. Not only can you configure standard and auto-proxy configurations, but you can configure custom proxy addresses for particular URLs. You can even assign a proxy to URLs that match a complex regular expression.

To set up these advanced configurations, you can use the Proxy Mapping feature of the Mobile Data Service included with BlackBerry Enterprise Server 4.0. Figure 8-2 shows the dialog that is used to set up a new HTTP Proxy Server mapping.

![New HTTP Proxy Server Mapping dialog]

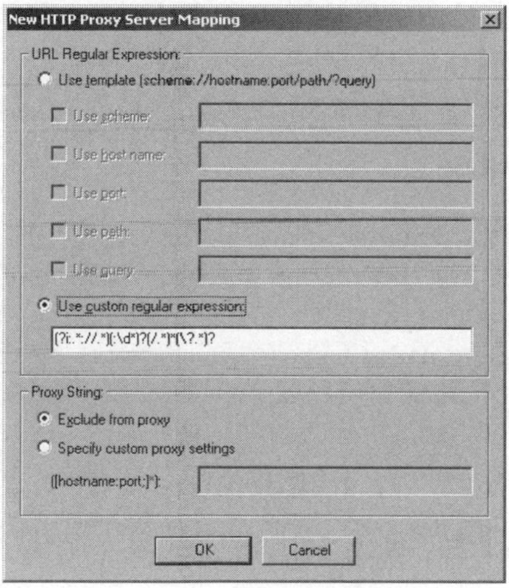

Figure 8-2. The New HTTP Proxy Server Mapping dialog

Select the Use template option to use a regular expression to match a specific portion of the URL or use the "Use custom regular expression" option to use the entire URL. Once you've added the regular expression, in the

Proxy String section, you can specify that the HTTP requests that match be excluded from a proxy or you can configure a custom proxy. This is where you would configure a bypass proxy list of URLs.

Hack the Hack

The capabilities in BES 4.0 provide full support for your bypass proxy lists and allow you to set up some pretty cool configurations. For example, you could create a configuration that disallows access to all URLs except those that you've explicitly allowed. Conversely, you could set up a disallow list for web sites to prevent access to certain URLs. You can configure MDS to send requests for those sites to a page that warns users that they've requested content that is deemed inappropriate.

Figure 8-3 shows a proxy mapping configuration that allows access to google.com, but other requests are proxied through *server.domain.com*. You could make a simple web page available at *server.domain.com* that gives the user a custom message.

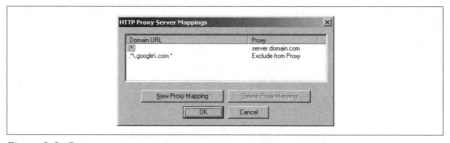

Figure 8-3. Custom proxy mappings

You could exclude all requests from a proxy except those that match your regular expressions. For the sites that match, you could set up a custom proxy that scrapes the content in the HTTP response and creates modified versions to send to the users. Check out *Spidering Hacks* (O'Reilly) for in-depth information on parsing HTML on the Web.

Track MDS HTTP Requests
#86
Figure out where your users are going using the BlackBerry Browser. You may even find some useful site that you want to visit.

Do your users use the Mobile Data Service? How frequently? Do your users go to Internet sites or intranet sites more often? Which sites are visited most?

All these questions are interesting ones that will be different in any environment. The more data you mine from the logs will translate into more knowledge about how your users are using the service and how their experience can be improved. Perhaps an intranet application that you know has not been optimized for the BlackBerry is nevertheless being visited but not fully utilized by your BlackBerry users. You may want to go to the application developer to push for better wireless support in the web application.

You might even see frequent access to various HTTP traffic for online games [Hack #32]. Perhaps you would want to disallow access to these and other non–business related sites.

A single HTTP request in your MDS logs takes up at least 16 lines, and the hostname and the rest of the URL are on different lines. This makes using a simple *grep* command to extract the URLs impossible. Here is some Perl code that keeps track of HTTP GET requests, reconstructs the original URLs and prints them by the number of requests made to each.

The Code

```perl
use strict;

# change the $mds_log variable to your log file
my $mds_log = "BESNAME_MDAT_01_20050607_0002.txt";

my $http_data = {};

open MDS, $mds_log;
while (<MDS>) {
    chomp;
    my $id = "";
    my $host = "";
    my $get = "";
    if (/EVENT = ReceivedFromDevice/ and
        /HTTPTRANSMISSION = Host:([^>]+)>/) {
            $host = $1;
          ($id) = (/CONNECTIONID = (\d+),/);
            $http_data->{$id}->{host} = $host;
    } elsif (/EVENT = SentToServer/ and
        /HTTPTRANSMISSION = GET (.+) HTTP\/1\.1>/) {
            $get = $1;
          ($id) = (/CONNECTIONID = (\d+),/);
            $http_data->{$id}->{get} = $get;
    }
}
close MDS;

my %urls = ( );
foreach my $id (keys %{ $http_data }) {
    my $host = $http_data->{$id}->{host};
    my $get = $http_data->{$id}->{get};
    next if not $host and $get;
    my $url = "http://$host" . $get;
    $urls{$url}++;
}
```

```
foreach my $url (sort { $urls{$b} <=> $urls{$a} } keys %urls) {
    print "$urls{$url}\t$url\n";
}
```

Your MDS logs appear with your other BlackBerry logs in the following directory where *YYYYMMDD* is the current date (assuming you've installed your BES on the *C:* drive):

```
C:\Program Files\Research In Motion\BlackBerry Enterprise Server\Logs\YYYYMMDD\
```

The Mobile Data Service logs will start with your BES name followed by the string MDAT.

```
BESNAME_MDAT_01_20050607_0002.txt
```

Type the code into your text editor and save it as *site_extract.pl*.

Run the Code

Bring up a command prompt, change directory (*cd*) to the directory where you saved the file, and type the following command to run the script:

```
C:>perl site_extract.pl
```

You'll get a list of URLs preceded by the number of instances they appear in your logfile, as shown in the following:

```
4    http://www.cnn.com/
3    http://dave.runningland.com/index.php
1    http://www.google.com/
```

You could easily pipe the output to a file and open the data using a spreadsheet program such as Microsoft Excel.

```
C:>perl site_extract.pl > todays.frequent.sites.txt
```

HACK #87 Detect BlackBerry Browser Requests the Right Way

Examine the headers of an HTTP request to determine what content is best suited for the browser making the request.

Let's say you've decided to have different versions of the content on your web site for desktop browsers than for the BlackBerry Browser. What is the best way to make these versions available to your users? You could make the different versions available at different URLs. For example, *http://www.site.com* would be for regular browsers while *http://wap.site.com* would be your URL for WAP browsers. While this approach certainly works, it forces your users to remember two different URLs for essentially the same content. It would be more convenient to use a single URL for your site and have your web application determine the content that should be provided, depending on the browser making the request. How can you check to see what type of browser is making a request to your web application?

There is a right way and wrong (read: easy) way to do this in your application. The good news is that the wrong way is okay to do in *certain situations*. There are two different HTTP headers that are available to your web application to help determine the browser type: HTTP_USER_AGENT and HTTP_ACCEPT.

The Old Way

The HTTP_USER_AGENT, or *user agent* string, is used by browsers as a way of advertising their browser type. This string has information about that identifies the browser and version along with general details about your operating system. For example, the following string is what is presented to web servers by Firefox on my Mac:

```
Mozilla/5.0 (Macintosh; U; PPC Mac OS X Mach-O; en-US; rv:1.7.7) Gecko/20050414
Firefox/1.0.3
```

The traditional method of determining browser type is parsing this value for certain identifying strings, such as MSIE 6.0 or Netscape 4.73. Back in the days when web browsing was reaching the tipping point in the late 1990s, this type of detection of browser types was commonplace and even required in some circumstances because of the differing behavior of the major browsers. Each browser supported a unique subset of HTML tags and handled some tags differently than other browsers, so web developers worked around these idiosyncrasies by parsing the user agent and returning different versions *by browser*.

This approach could certainly be used to serve up appropriate content to BlackBerry Browsers by parsing the user agent string for the word "Black-Berry." Most web servers provide access to this field through an environment variable, so a simple Perl regular expression for doing this would be:

```
if ($ENV{HTTP_USER_AGENT} =~ /BlackBerry/i) {
    # we've got a request from a BlackBerry
} else {
    # regular browser
}
```

This method is fine if you're certain that you only want to support Black-Berry's handheld browser. The problem with this approach is that there are other handheld browsers that you probably want to support as well. When you decide you want to support another handheld's browser, you'll have to change the code in your web application to support it! When another handheld comes along, you'll have to make another change. Before long, your browser detection code in your web application is growing large and complex. As a software developer, you want to keep your code as small and efficient as possible.

A Better Way

A more effective long-term solution is to interpret the HTTP_ACCEPT header, also presented by a web browser with every HTTP request. It specifies what media types the browser is capable of interpreting as a response from the web server. The HTTP_ACCEPT header is a comma-delimited list of MIME types along with some optional parameters specified with each. For example, here is the HTTP_ACCEPT value that Firefox presents to a web server:

```
text/xml,application/xml,application/xhtml+xml,text/html;q=0.9,text/plain;
    q=0.8,image/png,*/*;q=0.5
```

According to the HTTP specification, this HTTP_ACCEPT string means that the browser is capable of interpreting all the MIME types listed. In the case of text/html;q=0.9, the value of q is a way for the browser to indicate its preference if the server is capable of serving different types of content. Looking at all the text type in the HTTP_ACCEPT header, you can see that the browser is saying that it is capable of supporting content of type text; however, it prefers text of the XML variety since a nonexistent q parameter implies a q-value of 1. If text/xml is not available, then HTML is preferred (text/html;q=0.9), followed by plain text (text/plain;q=0.8).

If you have content tailored to a WAP browser, you can look for the following string to verify that the browser is capable of interpreting it: text/vnd.wap.wml. Look for the text/vnd.wap.wmlscript MIME type for browsers that support WMLScript (or at least claim to be able to).

By interpreting the HTTP_ACCEPT header and looking for a specific MIME type, you are not trying to guess the best content by first determining the browser type, but you're letting the browser tell you what content it handles best. This is a superior approach because you won't have to change your code when the next handheld browser comes along that you need to be able to support. That new browser will use the HTTP_ACCEPT header properly, and you'll already have it covered in your code.

This is the "proper" way for servers and web applications to do what is, in effect, browser sniffing. This technique wasn't possible a few years ago because of the browser shenanigans brought about by the browser wars between Microsoft's Internet Explorer and Netscape. Your web app could use the HTTP_ACCEPT header, but you still had to implement some way of determining the browser type because of the various quirks inherent with each browser. However, today's modern web browsers are fulfilling the previously unmet promise of a browser-agnostic web due in no small part to the browsers' support of Cascading Style Sheets [Hack #88].

See Also

- HTTP Specification *http://www.w3.org/Protocols/rfc2616/rfc2616-sec14. html#sec14.1*

HACK #88 Make Your Web Sites BlackBerry Friendly

Use these guidelines to make changes to your site to make it look great on a BlackBerry.

There are many web sites that will look surprisingly good on your Black-Berry without any modification at all. Others may require a tweak or two to make the content available to mobile devices. The good news is that you can make your site look good on a BlackBerry without sacrificing the user experience of your desktop browser–only visitors. There are certain guidelines that you can easily incorporate into your web development so that when you design sites, they'll look great for any device.

Don't Use Frames

Frames and mobile browsers just don't get along very well. It's easy to understand why: frames are made for very large screens. Because of this fact, the BlackBerry Browser doesn't even attempt to render frames—it forces users to choose which frame they'd like to view, one at a time. This is the bottom of the barrel when it comes to usability. Frames are often used as navigation on sites to display links to general areas on the site. As you can see from Figure 8-4, BlackBerry users will have little or no clue about which frame to use. The site displayed uses a left and a right frame.

Leaving BlackBerry usability aside, the sites that use frames well and make them usable even for desktop browsers are few and far between (Bloglines [Hack #39] is one; *http://www.bloglines.com*). Just try bookmarking a set of frames—it's not possible. Everything that frames are meant to accomplish

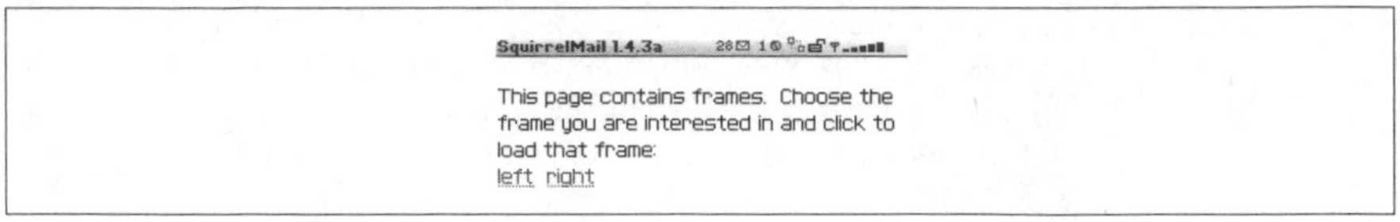

Figure 8-4. Frames in the BlackBerry Browser

can be done using other techniques in a better and more usable way. Couple this with the fact that they are almost impossible to use on a mobile browser, and frames are simply not a good design tool.

Limit Your Use of JavaScript

One of the limitations that the BlackBerry Browser has traditionally had is its lack of JavaScript support. Many sites are unusable on the BlackBerry because they require the client to support JavaScript and to have it enabled. This is not always the case even for desktop browsers. Requiring a browser to use JavaScript to simply navigate your links is bad design. The browser included with BlackBerry Handheld Version 4.0 does include some support for basic JavaScript—but this doesn't mean you should design your site requiring the use of every JavaScript method in the specification. The JavaScript implementation in 4.0 supports a subset of the methods being forced upon browsers by JavaScript-happy sites. Make sure JavaScript is optional or make a version of your site where it's optional and make it easy for your visitors to get to.

Use Images Wisely

Be very careful with the images you place on your web site. A common practice is to create an image with words on it and then use it as a link for navigation on your site. This was done in the early days of the Web so that web designers could be sure a site looked consistent across platforms and browsers without worrying about which fonts were installed or how a browser would display certain markup.

Many BlackBerry users choose to turn off images entirely [Hack #11], significantly speeding up their mobile web experience. These users won't even see the words on your images. Requiring the use of images for your site almost guarantees a slower experience for mobile users. While the time it takes to load images in a desktop browser is almost negligible in these current times of broadband access, your BlackBerry visitors will notice a huge lag.

If you must include images, be sure they are narrow enough to fit on a BlackBerry screen in a usable way. If you place wide images on the top of your page, your users will have to scroll down just to view the start of the actual content of your site. How rude!

Be sure and include alt text in your images as well. The alt text shows up on image placeholders in the BlackBerry Browser when users don't choose to display the full image (see Figure 8-5). Of course, I'm sure that all who read this have always followed the HTML specification and have never forgotten to include useful alternate text for every image on their sites.

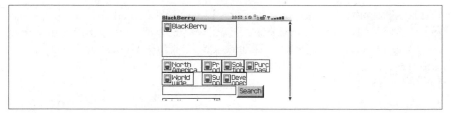

Figure 8-5. BlackBerry.com with images turned off

The most usable sites for mobile users and desktop users include no images in the HTML, but only in the Cascading Style Sheet (CSS). This allows your desktop and mobile browsers to view appropriate versions of the content without redirection or other tomfoolery.

Don't Use Tables for Layout

Using tables as a placement method is so early 2000s. It's time to join the revolt against this antiquated and expensive method of design and use only CSS for the layout of your site. Like frames, using tables for layout is almost always targeted at the desktop browser and shows neglect for the mobile browser. Move to a CSS-based layout, and you'll have almost automatic support for mobile browsers.

Test It Out

You don't need one of every BlackBerry made to test out your mobile site. In fact you don't need any. The BlackBerry Simulator **[Hack #93]** will do nicely.

Make BlackBerry-Friendly Sites from Scratch

We've discussed ways to modify an existing site to allow access from Black-Berry devices, but what guidelines should you go by to design a new site from scratch? Modern desktop web browsers like Firefox and Internet Explorer 6 are ushering in a new era of web design that promises a "code once, view anywhere" approach that allows a web site to be viewed unaltered on any web browser—including handheld browsers.

XHTML and CSS. The XHTML document type is garnering widespread support among major handheld browsers. While the format does have stricter requirements (all tags must be closed, all `<img>` tags must have `alt` attributes, etc.), you'll become a better web designer simply by using the format. Using XHTML for your documents' structure along with Cascading Style Sheets for the presentation will result in a more usable web site that looks more consistent across desktop browsers, and they'll look great on mobile browsers.

Use tags properly. In the past, we've been encouraged to write poorly designed HTML because the WYSIWYG web editing tools actually enforced bad design! Web developers were encouraged to use `<font>` tags to style text and use paragraph tags without closing them. And the browsers of old simply went along with it.

When you design your next site, step back and use the XHTML tag set properly. Plan your headings ahead of time. What headings should be between `<h1>` or `<h2>` tags? Why? If you plan your document structure ahead of time instead of simply typing away in your WYSIWYG editor, you'll save yourself a lot of CSS code later. Proper use of tags will make your document look great on handheld browsers without any styling at all.

Avoid deeply nested tables like the plague. From the beginning, plan your site's structure to have maximum flexibility in its presentation. CSS should be used for the entire layout—try to avoid table-based layout. If you insist on using tables for presentation, use them lightly. Don't use nested tables for layout purposes. The benefits of using a CSS layout are immense. If you use CSS for your layout, the next time you go through a redesign of your site, you won't have to change one element of your markup—just simply link to a new CSS file that contains your new design. Spend a few minutes at the *CSS Zen Garden* (*http://www.csszengarden.com*) to discover the power of this approach.

Make your pages small. The data networks on which the BlackBerry and other mobile devices operate, while improving, are still very slow compared to broadband. You should make your pages as light as possible so they'll load quickly on handheld browsers as well as desktop browsers. Separate the presentation from your XHTML code by placing the CSS in a separate file and then point to it with a `<link>` tag. This will reduce your page size (and bandwidth costs!) by large percentages. In addition, your CSS file will be cached in clients' browsers when they first visit your site so it won't have to be redownloaded for each page visited. Do the same with any JavaScript on your site.

See Also

- *Designing With Web Standards* by Jeffrey Zeldman (New Riders)

- CSS Zen Garden (*http://www.csszengarden.com*)
- "Why Tables for Layout Is Stupid" (*http://www.hotdesign.com/seybold/*)

HACK #89 Browse WML from Your Desktop

There are a couple options for viewing WML content from a non-handheld device (that is, your computer).

Once you've decided to take the plunge and create a handheld-only version of your site in the Wireless Markup Language (WML) format, you will quickly realize you've taken for granted the comforts of web development on your desktop computer. First of all, you can't view the source of a page on the BlackBerry Browser. View source is a very useful feature of modern desktop browsers that, as of yet, cannot be done on a BlackBerry. Equally as bad is when you are testing a change you made on your site and then refresh the page on your handheld to see the results—it takes ages to retrieve the document across the cellular data network compared to the same routine on your desktop. I won't even go into the convenience that the Web Developer Firefox extension brings to web development junkies.

You could use the BlackBerry Simulator [Hack #93] to view your site, but you still don't have access to viewing the source without coding your own program to do it. When I first dabbled in WML development, I wrote a Perl script that emulated an HTTP request from the BlackBerry Browser to view the source of a WML page—I sure wish I had access to the information in this hack back then!

Use Opera

The Opera web browser has a handler for WML that is turned on by default! This makes Opera a logical choice for developing WML sites. You can download Opera from *http://www.opera.com*. There are versions of the Opera web browser for Windows, Linux, Mac—almost any operating system you'd probably be using. The installation is quick and simple. Opera is free; however, ads appear in the toolbar. You can spend a few dollars to support Opera and get rid of the ads in your browser.

Once installed, you can visit any WML site and view it right in your browser. Figure 8-6 shows Opera 8.0 viewing a test WML page.

You can view the source of a WML page you are viewing by choosing Source from the View menu. Opera also gives you a quick way to access what it calls *Small Screen Mode*. This feature allows you to view an HTML site in a very small window, about the size of what you would see on a

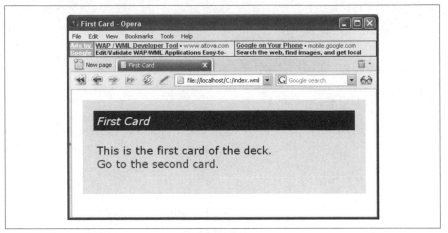

Figure 8-6. Viewing a sample WML page in Opera

BlackBerry. This mode can be accessed either by choosing Small Screen Mode from the View menu or by typing the keyboard shortcut, Shift-F11.

Opera treats local files that have *.wml* extensions as WML files—most browsers require the WML MIME type be returned in the HTTP response header to treat documents as WML. This is quite convenient if you are used to the "develop locally" approach that many web developers are accustomed to.

Use the Firefox wmlbrowser Extension

The ever-popular Firefox browser doesn't support WML content by default, but there is an extension that affords some of the functionality that Opera provides. The wmlbrowser extension for Firefox can be found at *http://wmlbrowser.mozdev.org/* and is installed just like any other Firefox extension. Detailed installation instructions can be found at this URL: *http://wmlbrowser.mozdev.org/installation/wmlbrowser.html*.

The wmlbrowser extension just adds a handler for the text/vnd.wap.wml MIME type that is returned in the HTTP response header on WAP sites. This allows you to view WML content from within Firefox. Figure 8-7 shows a view of the same WML page as before using the wmlbrowser extension. Notice the different appearance of the page compared to the rendering by Opera. The wmlbrowser extension shows all the cards in your deck while Opera shows one card at a time, which is similar to the behavior of most "real" WAP browsers.

Just as in Opera, the wmlbrowser extension allows you to view the source of a WML page just as you would for a page that's formatted in HTML. Unfor-

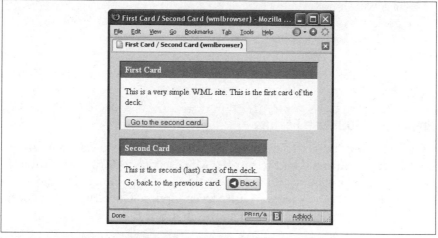

Figure 8-7. Viewing a WML page using the wmlbrowser extension

tunately, there is no syntax highlighting for WML like there is for HTML source in Firefox.

Because the wmlbrowser simply adds some code to handle the WML MIME type, you also cannot view local files and have them rendered as WML. The flexibility Opera provides makes it a better option for WML development at this time. Neither wmlbrowser nor Opera can interpret WMLScript at the time of this writing.

HACK #90 Create a Simple Push Application

You don't have to be able to write J2ME code to create a useful BlackBerry application. The BlackBerry Browser along with MDS allows you to push data to devices.

As ubiquitous as Java is now, not every programmer is in love with the language. But since the BlackBerry is a J2ME-based device, if you want to create a corporate based application, you have to use Java, right? Actually, no. RIM provides excellent hooks into the BlackBerry Browser so that when used in conjunction with a BES, you can create much of the same functionality of a full-blown J2ME application in any language that can use HTTP. You can create applications that push web content to certain devices so that it is there without the user having to pull the content by visiting the page in the BlackBerry Browser. The content that is pushed is cached on the device, making it available whether the user is in a wireless coverage area or not. The Mobile Data Service provides flexibility in how you notify the user that new content has been pushed to the device.

This hack will show you how to create a simple push application to give you a jumpstart in the world of MDS Push. You can even test this using the BlackBerry Simulator [Hack #93].

> For this to be a viable solution, your target audience must have devices homed on your BES infrastructure.

Enable the MDS Push Feature

By default, the Mobile Data Server is not enabled for push applications. You'll need to enable it and restart the MDS service to turn it on. In BlackBerry Manager, right-click on the server that you'd like to enable and select the "Set as Mobile Data Service Push Server" option. Once enabled, a checkmark appears before the option, as shown in Figure 8-8.

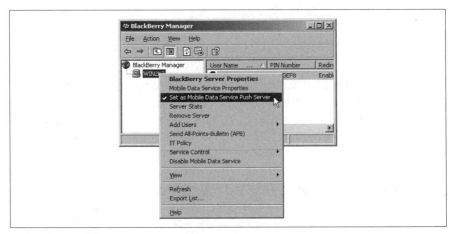

Figure 8-8. Enabling the push server

Design Your Push Application

You'll need a couple pieces of data to send content to a device. Here is a list of the minimum:

- Name of MDS push-enabled BES
- Port of MDS web server (default is 8080)
- Either the PIN or email of the recipient

Once you enable the MDS Push Server, a version of Apache Tomcat is made available on the BES to accept HTTP requests from your application. To push content to a device, you'll need to use an HTTP POST to a specific URL on your MDS server. You'll specify the device information in the URL parameters. There are several ways to customize the behavior of your push application through the use of custom HTTP headers in the request. Table 8-1 shows a list of the headers that are available and what functionality they provide.

Table 8-1. Custom HTTP headers for MDS Push applications

Header name	Description
X-RIM-Push-Type	Values can be Browser-Message, Browser-Content, or Browser-Channel (see description later in this hack).
X-RIM-Push-Title	Display name of your application. Displayed in the Home screen or message subject depending on value of X-RIM-Push-Type.
X-RIM-Push-Channel-ID	String that identifies your application.
Content-Location	The URL for the content you are sending.
X-RIM-Push-UnRead-Icon-URL	A URL for an image to display on the Home screen when your application has sent data that has not yet been read by the user.
X-RIM-Push-Read-Icon-URL	A URL for an image to display on the Home screen when your application has no new content on the device.
X-RIM-Push-Priority	Value can be none (default), low, medium, or high, each with increasing levels of intrusiveness to the user.
X-RIM-Push-Ribbon-Position	Specifies where your application icon appears in relation to the other icons on the Home screen.
X-RIM-Push-Description	A brief description of your application.
X-RIM-Push-Deliver-Before	Specifies a date that the content must be delivered by. If it is not sent by the specified date, MDS will not push the data.
X-RIM-Push-ID	Specifies a unique message ID that can be used to cancel or check status of delivery.
Content-Type	Defines the MIME types included in the pushed content.
Cache-Control	Value can be no-cache, max-age, or must-revalidate.
X-RIM-Transcode-Content	Specifies which type of content the server should transcode (convert from one digital format to another).
X-RIM-Push-Reliability	Specifies the delivery reliability mode of the content. Value can be Transport, Application, or Application-Preferred.
X-RIM-Push-NotifyURL	Specifies a URL for the MDS server to send a result notification.

Push Type

The X-RIM-Push-Type header determines the location and appearance of the application on the BlackBerry. There are three types of pushes you can make to the device:

Browser-Channel

>   Creates an icon on the Home screen. You can create custom icons that display when your application contains read or unread data. See Figure 8-9.

Browser-Message

>   New content shows up in the message list with a special icon that differentiates it from regular messages. See Figure 8-10.

Browser-Content

>   Sends content to device where it's stored in the browser cache. There is no icon that represents new content.

Figure 8-9. Browser-Channel push app on Home screen

Figure 8-10. Browser-Message push app in message list

The Code

```
use strict;
use LWP::UserAgent;
use HTTP::Request;

my $MDS = "localhost";
my $PORT = "8080";
my $PIN = '2100000A';
my $url = "http://del.icio.us/html/davemabe/?tags=no&count=20";
my %headers_to_send = (
 'Content-Location'      => $url,
 'Content-Type'         => "text/html",
 'X-RIM-Push-Title'      => "Dave's del.icio.us Bookmarks",
 'X-RIM-Push-Type'       => "Browser-Channel",
 'X-RIM-Push-Channel-ID' => "daves-delicious",
);

my $browser = LWP::UserAgent->new;
```

```
my $get_response = $browser->get($url);
my $content = $get_response->content;

my $mds_url =
  "http://$MDS:$PORT/push?DESTINATION=$PIN&PORT=7874&REQUESTURI=/";

my $request = HTTP::Request->new;
$request->method('GET');
$request->uri($mds_url);

foreach my $header (keys %headers_to_send) {
  my $value = $headers_to_send{$header};
  $request->header($header,$value);
}

$request->content($content);

my $response = $browser->request($request);

if ($response->is_success) {
  print "Push was successful.  Sent to $PIN.\n";
} else {
  print "There was a problem.  Code was: ",$response->code,"\n";
}
```

Run the Code

Type the code into your favorite text editor and save the file as *mdspush.pl*.
Bring up a command prompt and type the following in the directory where
you saved the file.

```
C:>perl mdspush.pl
```

Output

You'll get different output from this command depending on whether your
push was successful or not. If the push was successful, you will get the following output:

```
Push was successful.  Sent to 2100000A.
```

If there was a problem with the request, you will get an error message along
with the HTTP response code that MDS returned:

```
There was a problem.  Code was: 500
```

In the event the code is unsuccessful, there are some common errors that
should be checked. In my testing, I created some errors and got these status
codes. An HTTP response code of 500 was given when I used an incorrect URL
for my MDS. A response code of 403 indicated I had either sent to the wrong
PIN, or the URL for the content I was sending to the device was invalid.

When your code is successful, you should get an icon on the device's Home screen, as shown in Figure 8-9, and when you click on the icon, you should see a list of links I've recently posted on my del.icio.us page, as shown in Figure 8-11.

Figure 8-11. The pushed content on the device

As well as placing an icon on the Home screen, a browser channel push will add an additional folder underneath your BlackBerry Browser's bookmarks that contains all channel applications on the device. The user can use the trackwheel to open, view details, reorder, or delete the channels you push.

HACK #91 Delete a Push Application

Once a push application is no longer needed, be courteous and clean up after yourself. Don't litter—send the unused icon to the bit bucket!

We all would like to create useful, elegant software applications that users love and that last forever. Usually it doesn't turn out so rosy. There are bug fixes, new features, and upgrades that are generally required. There are some applications that simply outlive their usefulness.

Once your application is due for a major upgrade or you need to send your program "out to pasture," it's a good idea to clean up after yourself. Once your users start installing some of the third-party applications in this book, their Home screens (or Applications submenus on 7100 series devices) will quickly become used-car lots of icons. You don't want to add to the junk pile. Use this code to push a command that deletes your push application.

Identify Your Application

When you pushed your application for the first time, in addition to a display name, you included a string that uniquely identifies your application on the user's handheld using the custom X-RIM-Push-Channel-ID HTTP header. This string is not seen by the user, but the device knows to use this string as the program's identity. When new content is pushed to the BlackBerry, the device checks whether it has already received content for the application and created a *channel* for it. If this channel has already been created, the device updates the channel with your new content. If your application's ID doesn't exist on the device, it sets up a new channel for your program.

This same string is used when you need to delete your application (or channel) from a device.

The Code

The code for deleting your application looks much the same as the code for creating one [Hack #90]. This code deletes the push application that was created in that hack that sent the latest links from my del.icio.us page [Hack #46].

```
use strict;
use LWP::UserAgent;
use HTTP::Request;

my $MDS = "localhost";
my $PORT = "8080";
my $PIN = '2100000A';

my %headers_to_send = (
    'X-RIM-Push-Type'        => "Browser-Channel-Delete",
    'X-RIM-Push-Channel-ID'  => "daves-delicious",
);

my $mds_url =
  "http://$MDS:$PORT/push?DESTINATION=$PIN&PORT=7874&REQUESTURI=/";

my $request = HTTP::Request->new;
$request->method('GET');
$request->uri($mds_url);

foreach my $header (keys %headers_to_send) {
    my $value = $headers_to_send{$header};
    $request->header($header,$value);
}

my $content = "this string can be anything";
$request->content($content);

my $browser = LWP::UserAgent->new;
$browser->agent("My Test Push Application");

my $response = $browser->request($request);

if ($response->is_success) {
    print "Delete was successful.  Sent to $PIN.\n";
} else {
    print "There was a problem.  Code was: ",$response->code,"\n";
}
```

Running the Code

Type the previous code in your favorite text editor and save the file as *pushdelete.pl*. Bring up a command prompt and type the following in the directory where you saved the file.

```
C:>perl pushdelete.pl
```

Output

You'll get different output from this command depending on whether your push was successful or not. If the push was successful, you will get the following output:

```
Push was successful.  Sent to 2100000A.
```

If there was a problem with the request, you will get an error message along with the HTTP response code that MDS returned:

```
There was a problem.  Code was: 500
```

Push Install Applications

#92

If there is an application that you'd like all the handhelds on your BES to install, you can push it out silently over the air.

Whether you have developed a custom application [Hack #90] in-house or you'd like to make a killer application available to all your users, you can do a silent install for users over the air. This is a great way to install applications on behalf of your users, saving them what could be a confusing task for some inexperienced users. The handhelds you're pushing to will need to already be running 4.0 and your BlackBerry Enterprise Server needs to be as well.

Prepare the Application

You'll need to create a Windows share on a computer that has Desktop Manager installed. You could use your BES server, but you don't have to. Once you install Desktop Manager on your selected machine, you'll need to share the following directory (the actual share name is not important—the default works fine): *C:\Program Files\Common Files\Research In Motion*.

Next, create a folder called *Applications* in the directory *C:\Program Files\ Common Files\Research In Motion\Shared*. Inside the *Applications* folder, create a directory for each application that you'd like to install, Berry 411 [Hack #51], for example. Put the *.alx* and *.cod* files for your application inside this directory. You might need to modify the *.alx* file and delete any `<directory>` elements in the file. Once you have your application files in the appropriate directory, open a command prompt, and change directory to *C:\Program Files\Common Files\ Research In Motion\AppLoader,* and run the command **loader /reindex**.

Prepare the BES

On your BlackBerry Enterprise Server, bring up the BlackBerry Handheld Configuration Tool by going to your Start menu, and then Program Files → Research In Motion → BlackBerry Handheld Configuration Tool. Select Software Configurations on the left side. Click on the "Add New Configuration" task in the bottom-right pane. Select a name for your configuration and type a description. In the Handheld Software Location field, click Browse and type the share name that you created in the previous section. If you've properly configured the share and your *.alx* and *.cod* files, your applications will show up as you expand *Applications*, as shown in Figure 8-12.

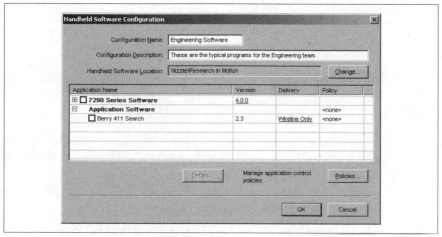

Figure 8-12. Creating a new Handheld Software Configuration

> The idea of software configurations in 4.0 is to create a configuration and then assign it to several handhelds for users that have like needs. It's a good idea to create software configurations with groups of people in mind, like "Engineering Software" or "Executive Tools."

Check the box next to each software application that you'd like to push to your users. Under the Delivery column, select "Wireless" to deliver the program over the air or "Wireline Only" to have the application installed the next time the user runs Application Loader on her computer.

Click the Policies... button on the bottom to configure a new policy for this program. Click the New... button to create a new policy. Choose a descriptive name for the policy in the Name field. In the Disposition field, be sure to select Required to make the installation of this program mandatory (see Figure 8-13).

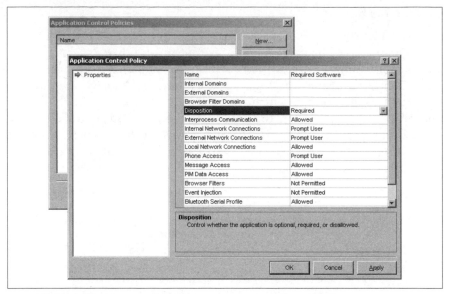

Figure 8-13. Adding a new policy

Click OK twice to return to your new software configuration and select your new policy from the list in the Policy column, and click OK, as shown in Figure 8-14.

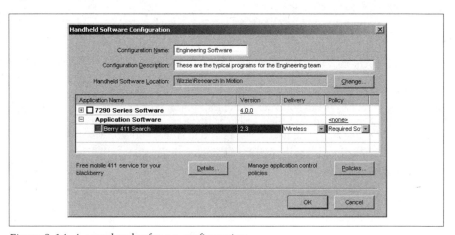

Figure 8-14. A completed software configuration

Assign the Policy to a Handheld

Once you've created a software configuration, you can assign it to any handheld with 4.0 on your BES. Click on Handhelds on the left pane, and then select a handheld in the right pane to send the software to. Select the Assign

Software Configuration task and then select your newly created software configuration from the list and click OK, as shown in Figure 8-15.

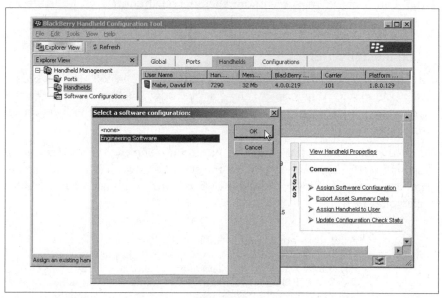

Figure 8-15. Assigning the software configuration

On the next push interval, the software will appear on the handheld you've selected. The push interval is about four hours, so be prepared to wait a while to see your results. You can monitor the BlackBerry Policy Service log-files (with the "POLC" identifier by default) on your BES to see exactly what it's doing with your new policy. You can use the following command to filter your logs for the entries associated with pushing applications.

```
C:\Your Log File Directory\20050620\>findstr SendApp *
```

HACK #93   Simulate Any BlackBerry

Is that image too wide for a BlackBerry screen? Is it going to look the same on a 7100g as it does on the 7290 attached to your hip? Use the simulator to find out.

Wouldn't it be nice if you could get each and every BlackBerry device that is released to put it through the paces? Does the entire web development staff need a BlackBerry solely to see how their work appears on a device? I think your money is more wisely spent in other ways and so does RIM.

The BlackBerry Simulator is not just a tool for hardcore application developers that are holed up in a dark room with black poster board over the windows. It comes in quite handy for web developers as well—even your help desk could find some use for it. It comes with a slimmed down version of the Mobile Data Service, so you can get extremely close to the actual end user experience without shelling out hundreds of dollars for a device and service.

Obtain the Software

To use the simulators, you'll need to download and install the BlackBerry Java Development Environment. It is a free download, however, you will have to cough up some personal information and agree to a lengthy end user license agreement (EULA) to use it.

Go to the BlackBerry JDE web site *http://www.blackberry.com/developers/ downloads/jde/index.shtml* and follow the steps to download the 4.0 version of the software. The download file is over 60 MB, so go get a cup of coffee while you wait.

 As of this writing, Version 4.0.1 of the JDE is available. This version is actually an update to Version 4.0, so you'll need to have already installed 4.0. The only updates in 4.0.1 are the location API libraries for the new GPS devices. If you don't have a need for the GPS libraries, it's safe to skip the 4.0.1 release.

The JDE is installed much the same as any Windows application. Make sure you choose the complete installation to get all the goodies.

Run the Simulator

Once you've installed the BlackBerry JDE, bring up a command prompt and *cd* (change directory) to the directory where the device simulators are located. By default, they are located in *C:\Program Files\Research In Motion\ BlackBerry JDE 4.0\simulator*.

In that directory, there will be several files including batch (*.bat*) and XML configuration files for each device that the JDE can simulate. Run the batch file for the BlackBerry model of your choice.

```
C:\Program Files\Research In Motion\BlackBerry JDE 4.0\simulator> 7290.bat
```

After a moment or two, you should see the device simulator appear as simply another application on your desktop (see Figure 8-16).

Access the Keys

Most of the keys will map to their corresponding keys on your computer's QWERTY keyboard, but you'll have to take note of the ways to access the

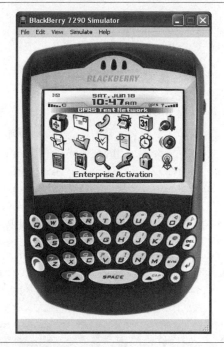

Figure 8-16. The 7290 simulator

keys that are unique to the BlackBerry (the trackwheel, for instance). Table 8-2 shows how the device keys map to the PC keyboard.

Table 8-2. Accessing the unique device keys

Key	Using keyboard	Using mouse
Alt	Control	Click the Alt key
Rolling the trackwheel	Up and down arrows	Use the mouse scrollwheel
Clicking the trackwheel	The Enter key or left arrow	Click the trackwheel
The Escape key	The Escape key or right arrow	Click the Escape key
Backlight	Page Down	Click the Backlight key
The Symbol key	Delete	Click the Symbol key
The Phone key	Page Up	Click the Phone key

Browse the Web

When you try to access a web site using the browser, you'll get an error message, as shown in Figure 8-17.

Figure 8-17. The error when the MDS simulator isn't started

To get the browser to work, you'll need to start the Mobile Data Service simulator. This program simulates an actual MDS instance on your desktop, allowing the browser to access web sites just as if it were accessing it using a real MDS. Go to your Start menu, and then select Program Files → Research In Motion → BlackBerry Java Development Environment 4.0 → MDS Simulator. This brings up a command window that stays running in the background, as shown in Figure 8-18. You can check this window for error messages and other activity as you do your testing. The MDS simulator and the email simulator described in the next section can be started in any order—before or after you start the handheld simulator.

Figure 8-18. The MDS simulator command window

Once the MDS simulator is running, you can use the BlackBerry Browser to test any web site that is accessible from your computer. Because it's an (almost) fully functioning Mobile Data Service, you can test your push applications using the simulator. The default PIN for the handheld simulator is 2100000A. This can be changed by editing the batch file used to start the specific model's simulator. Look for the /pin=0x2100000a string in the file and change it to use another PIN.

Simulate Email

The BlackBerry JDE also provides an email server simulator. This provides basic sending and receiving functionality to your handheld simulator. You can point the email simulator to a POP3 mailbox and have new messages delivered to your handheld. It also allows you to enter an SMTP server to deliver outbound messages from your device. Start the Email Server Simulator by going to your Start menu and selecting Program Files → Research In Motion → BlackBerry Java Development Environment 4.0 → Email Server Simulator. A command window will appear, and eventually a dialog appears that allows you to enter your POP3 configuration for your mailbox along with the outbound SMTP server (see Figure 8-19).

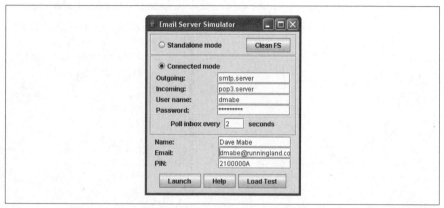

Figure 8-19. The email server simulator

If you try to send email messages on your handheld simulator without the email server simulator running, the messages will stay in the message list as if you were out of network coverage on a real BlackBerry device. Once you start the simulator, they'll be sent immediately, as if you came back into a coverage area.

Hack the Hack

When you peer into the batch files that start the simulator, you'll see that they call an executable named *fledge.exe* that actually runs the simulator. If you run *fledge.exe* directly, you can use a variety of command-line options to customize the simulator's behavior. Table 8-3 shows some of the interesting ones to try on for size.

Table 8-3. Command-line switches for the simulator

Command-line switch	Function
/help	Show all command-line switches.
/pin={int}	Give the device a pin of your choice.
/secure	Enable device security such that custom apps have to be signed [Hack #98].
/session={string}	Write to specific logfiles for this session of the simulator.
/title={string}	Use a custom title for the simulator window.
/no-show-plastics	Don't show the device, just its LCD screen.
/zoom={float}	Magnify the simulator by the given factor.
/no-save-flash	Don't save the flash memory to disk upon exit.
/phone-number={number}	Assign a custom phone number to the simulator.

CHAPTER NINE

Application Development
Hacks 94–100

Over the last several months, an explosion of new BlackBerry applications has hit the market. If there was ever any doubt about the viability of the BlackBerry as a development platform, it has been reduced to a whimper. There are applications for getting real-time stock quotes [Hack #70], spellchecking [Hack #65], even a bartending program.

If you speak a little Java, you can brew up your own application to communicate with web servers [Hack #94]. Best of all, there are no license fees to get started—in fact, you don't even need to own a BlackBerry device [Hack #93]. RIM provides free access to the BlackBerry JDE, a development kit that includes an IDE, or *integrated development environment*. The other nice feature of the BlackBerry platform is there are a variety of ways [Hack #97] for your users to install your program, even over the air wherever they happen to be. Use the hacks in this chapter to get started.

HACK #94 Create HTTP Connections
Create simple HTTP requests from a BlackBerry application.

The most powerful feature of programming BlackBerry applications is having the ability to make your applications perform wireless network connections to servers. The most common network connection type is HTTP allowing applications to connect to web servers anywhere on the Internet.

Create HTTP GET Requests

After importing the *java.io.* * and *javax.microedition.io.* * packages, here's is all you need in J2ME to create an HTTP connection:

```
HttpConnection c = (HttpConnection)Connector.open("http://www.yahoo.com");
```

This will perform an HTTP connection to the Yahoo! web server. To read the response from the server, the first thing you must do is read the HTTP response code:

```
int responseCode = c.getResponseCode( );
// check response code
```

The following step is to read the body of the HTTP response. To do this, you must open the InputStream of the response and read the stream until no more data is available:

```
InputStream in = c.openInputStream( );
int i = in.read( );
StringBuffer sb = new StringBuffer( );
while (i!=-1) {
    sb.append((char)i);
    i = in.read( );
}
```

One special requirement to perform network connections on the BlackBerry is that connections must be performed in a separate thread from the UI thread. For example, if the user clicks a menu item or presses a button, you cannot place your HTTP request code in the same thread as the event handler, you must spawn off a new thread. There are a couple of reasons for this: first, it prevents the BlackBerry UI from locking up while the request is in process; second, by default the BlackBerry will pop up a dialog box requesting that the user accept this network connection, as shown in Figure 9-1.

The user must select "Allow this connection" for the connection to succeed. If the user does not select to allow it, a security exception will be thrown to your application.

Here is some sample code that will spawn the HTTP connection to be performed in a separate thread:

```
(new Thread(new Runnable( ) {
    public void run( ) {
        HttpConnection c =
            (HttpConnection)Connector.open("http://www.google.com");
        c.getResponseCode( );
    }
})).start( );
```

Configure Connections to Use the MDS or Direct TCP/IP

One of the most unique and powerful features of the BlackBerry device is its ability to use a BlackBerry Enterprise Server (BES) to connect securely to an email server. Starting with BES 2.2 and above, there is a new feature called the Mobile Data Service (MDS), which allows applications to perform network connections through the BES server. You can think of the MDS as a VPN

Figure 9-1. Allow HTTP connection dialog

server to which the BlackBerry is permanently connected. All the data sent between the BlackBerry device and the MDS is encrypted exactly the same way as emails are encrypted by the BES. This means that your application can make connections to servers inside a private network without any ports needing to be opened on the firewall, since all connections are tunneled through the MDS server located inside the firewall.

Another method of performing network connections from the BlackBerry device is by using the built-in TCP/IP stack. The TCP/IP stack allows the BlackBerry to make network connections using the carrier's network connection to the Internet, whether the device is attached to a BES server or not. To make TCP/IP connections, the device must have Handheld System Software 4.0 or above installed.

By default, most BlackBerry devices attempt to create their network connections using the MDS. There is one exception: iDEN BlackBerry devices make TCP/IP connections by default.

It is possible to make a special modification to your code to force a connection to use either the MDS or the TCP/IP stack:

```
// this code will force the use of MDS
Connector.openConnection("http://www.google.com;deviceside=false");

// this code will force to use the built in TCP/IP stack
Connector.openConnection("http://www.google.com;deviceside=true");
```

—Paul Dumais

H A C K
#95 Create a Simple Stock Quote Application

Build a simple stock quote application that will prompt a user for a stock symbol, then perform an HTTP request to the Yahoo! stock quote server, and finally display the stock price.

To create BlackBerry applications, you must download and install the Black-Berry Java Development Environment (JDE) from the *http://www.blackberry.com* web site. The JDE contains the development environment, compiler, MDS Simulator, and the BlackBerry Device simulator [Hack #93]. Once the JDE is installed, you are ready to start writing your application.

Launch the JDE, and create a new workspace and a project and call it whatever you want. Then add a new *.java* file to your project. The simplest version of a BlackBerry UI application must simply contain a public static void main(String[] args) method. Here is an example of the simplest BlackBerry application:

```
import net.rim.device.api.ui.*;
public class StockQuotes extends UIApplication {
    public static void main(String[] args) {
    }
}
```

This application does not do anything—it does not even display a user interface. The next step is to create a user interface.

For this stock quote example, we will simply show a text input field in which the user can enter the stock symbol:

```
import net.rim.device.api.ui.*;
import net.rim.device.api.ui.component.*;
import net.rim.device.api.ui.container.*;

public class StockQuotes extends UIApplication {
    public static void main(String[] args) {
        StockQuotes theApp = new StockQuotes();
        theApp.enterEventDispatcher();
    }
```

```
public StockQuotes( ) {
    MainScreen stockScreen = new MainScreen( );
    stockScreen.setTitle("Stock Quotes");
    stockScreen.add(new EditField("Symbol: ", "rim.to"));
    pushScreen(stockScreen);
}
}
```

When the application is run, a simple interface is displayed to the user, as shown in Figure 9-2.

Figure 9-2. Stock Quotes main screen

The next step is to add a menu item. When clicked, this menu item will perform an HTTP connection to the Yahoo! Finance stock quote server:

```
stockScreen.addMenuItem(new MenuItem("Get Quote!", 1, 1) {
    public void run( ) {
        // place event handler code here
    }
});
```

Now in the event handler code, simply add the necessary HTTP request code [Hack #94]. When the menu item is clicked, the stock quote is fetched and the result displays, as shown in Figure 9-3.

Figure 9-3. Stock Quotes result

The Code

```
import java.io.*;
import javax.microedition.io.*;
import net.rim.device.api.ui.*;
import net.rim.device.api.ui.component.*;
import net.rim.device.api.ui.container.*;

public class StockQuotes extends UiApplication {

    private EditField symbolField = new EditField("Symbol: ", "rimm");
    private RichTextField priceField =
      new RichTextField("", Field.READONLY);

    public static void main(String[] args) {
        StockQuotes theApp = new StockQuotes();
        theApp.enterEventDispatcher();
    }

    public StockQuotes() {
        MainScreen stockScreen = new MainScreen();
        stockScreen.setTitle("Stock Quotes");
        stockScreen.add(symbolField);
        stockScreen.add(priceField);
```

```
stockScreen.addMenuItem(new MenuItem("Get Quote!", 1, 1) {
    public void run( ) {
        (new Thread(new Runnable( ) {
            public void run( ) {
                getQuote( );
            }
        })).start( );
    }
});
pushScreen(stockScreen);
}

private void getQuote( ) {
    try {
        HttpConnection c = (HttpConnection)
            Connector.open("http://finance.yahoo.com/d/quotes.csv?s="
            + symbolField.getText( )+"&f=l1");
        c.getResponseCode( );

        InputStream in = c.openInputStream( );
        int i = in.read( );
        StringBuffer sb = new StringBuffer( );
        while (i!=-1) {
            sb.append((char)i);
            i = in.read( );
        }
        priceField.setText("The last ask price for '" +
                symbolField.getText( )+"'  was $"+sb.toString( ));
    }
    catch (IOException e) {
        priceField.setText("Error: "+e);
    }
}
}
```

Run the Code

Create a new project in the BlackBerry JDE called StockQuotes. Underneath the project, make a new file called *StockQuotes.java*. Save the previous text into the file, and then go to the Build menu and select Build All and Run. This will compile your code and start the simulator with the application already "installed" on the BlackBerry.

With HTTP requests and simple applications mastered, it is now possible to build any type of networked application such as web browsers or custom applications.

—Paul Dumais

HACK #96 Integrate the Browser into a Java App

Combine the richness of a Java client with the extensibility and power of a web backend.

Berry 411 is built using a Java client that launches a web browser. This is a powerful and convenient way to build mobile applications.

The Java client provides instant accessibility from the Home screen and a rich native BlackBerry UI. The browser results screen allows for zero-install deployment of new features, easy display of rich text and graphics, and linking to the expanding array of mobile-friendly web content.

The BlackBerry is especially well suited to building this sort of application. The Back button works as expected when going back from a browser page to the client application that launched it. Using the built-in browser network configuration avoids many of the configuration hassles of direct network access, which requires a number of carrier-specific parameters.

Launch the Browser from Java

In this hack, we'll build an application to do reverse phone number lookups, going from a phone number to a name and address. The Java client displays a titlebar and input field, filtered to limit the user's input to 10 digits. When the user selects Search from the menu (see Figure 9-4), the application launches a web page (in this case, an existing Infospace service) to display the results, shown in Figure 9-5.

Figure 9-4. Java client

Figure 9-5. Web results

Even in this very simple form, the hybrid approach has advantages over a pure browser solution. The application launches instantly and has a degree of control over the user's input, which could not be achieved in a browser. It would not be difficult to extend this application to take further advantage of the rich Java client; for example, it could allow the user to select a phone number from his Address Book to look up.

The Code

```
/**
 * Reverse Phone Directory Lookup Application
 * This sample uses a Java frontend to prompt the user for a phone
 * number, then launches a web page to display the corresponding address.
 */
import net.rim.device.api.ui.*;
import net.rim.device.api.ui.component.*;
import net.rim.device.api.ui.container.*;
import net.rim.device.api.i18n.*;
import net.rim.device.api.system.*;

import net.rim.device.api.collection.util.*;

public class BrowserLauncher extends UiApplication
{
    public static void main(String[] args)
    {
        BrowserLauncher theApp = new BrowserLauncher();
        theApp.enterEventDispatcher();
    }

    public BrowserLauncher()
    {
        pushScreen(new SearchScreen());
    }

    class SearchScreen extends MainScreen
    {
        BasicEditField _phoneNumberEntry;

        SearchScreen()
        {
            super(DEFAULT_MENU | DEFAULT_CLOSE);
            setTitle(new LabelField("Reverse phone lookup",
                LabelField.ELLIPSIS | LabelField.USE_ALL_WIDTH));

            _phoneNumberEntry = new BasicEditField("Phone number: ",
                "", 10,
                BasicEditField.NO_NEWLINE | BasicEditField.FILTER_INTEGER);
            add(_phoneNumberEntry);
            addMenuItem(new MenuItem("Search", 1, 1) {
                public void run() {
                    String phone = _phoneNumberEntry.getText();
                    launchBrowser("http://www.infospace.com/" +
                        "_1_L5UTTJO3LEBVO__infow.sbcw" +
                        "/x/app/rev/detail.xhtml?top=&area=" +
```

```
                              phone.substring(0,3) + "&exchange=" +
                              phone.substring(3,6) +
                              "&phend=" +  phone.substring(6,10) +"&ran=24");
                }
            });
        }
    }

    static private boolean launchBrowser(String baseUrl)
    {
        boolean retval = true;

        int handle =
          CodeModuleManager.getModuleHandle("net_rim_bb_browser_daemon");

        if (handle <=0 )  {
            retval = false;
        }
        else {
            ApplicationDescriptor[] browserDescriptors =
              CodeModuleManager.getApplicationDescriptors(handle);

            if (browserDescriptors == null )
            {
                retval = false;
            }
            else {
                if ( browserDescriptors.length <=0 ) {
                retval = false;
                }
                else {
                    String[] args = {"url", baseUrl};
                    ApplicationDescriptor descriptor =
                    new ApplicationDescriptor(browserDescriptors[0],
                        "url invocation", args,
                        null, -1, null, -1,
                        ApplicationDescriptor.FLAG_SYSTEM);
                    try {
                        ApplicationManager.getApplicationManager( ).
                            runApplication(descriptor);
                    }
                    catch(ApplicationManagerException e) {
                        retval = false;
                    }
                }
            }
        }
        return retval;
    }
}
```

Save the code as *BrowserLauncher.jav'a*: this is the sole class in this application. The key function is launchBrowser( ), which includes some black magic
to launch the user's default browser with the requested URL. The code links to
the Infospace reverse directory page, providing the user's input as parameters.

Compile and Run the Application

To compile this code, use the Research In Motion JDE, which you can
download from the BlackBerry web site. The code has been successfully
tested with Versions 3.7 and 4.0 of the JDE.

Create a new workspace and project in the JDE, and add *BrowserLauncher.
java* to the project. (If you'd like to save yourself the typing, you can download a ZIP archive, including the entire project, from *http://thebogles.com/
browserlauncher.zip*.)

Select "Go" from Debug menu, and the JDE will build the application and
launch the BlackBerry simulator, including the application. You should also
launch the MDS simulator from the JDE Start menu item to allow the application to access the network.

Once the simulator is launched, you can scroll to the BrowserLauncher
application on the Home screen and click on it, by using your mouse scroll-
wheel. Type in a listed 10-digit phone number, click on "Search in the
menu, and you should see the results displayed in the browser.

Sign and Distribute the Application

Launching the browser is considered a "controlled API" by Research In
Motion. To run the application on a real BlackBerry, you will also need to
register for a code signing key and sign the application [Hack #98].

—Phil Bogle

Deploy BlackBerry Applications

Here are three ways to package your applications to be easily deployed on
BlackBerry handhelds.

One of the most important things to consider when planning the deployment
of a BlackBerry application is to determine how your customers will transfer
the application onto their handheld devices. For best results, it is highly recommended that you offer your customers as many options as possible.

BlackBerry applications are developed in the Java Development Environment (JDE) [Hack #93] and produce a COD file. However, you need more than
a COD file to successfully deploy the application to a handheld.

This hack will discuss how to package applications for deployment through the BlackBerry Desktop Manager, for over-the-air deployment, and to be pushed onto the device.

Deploy via the BlackBerry Desktop Manager

The BlackBerry Desktop Manager was the first method developed to deploy third-party applications onto BlackBerry handhelds. This method allows you to load applications from a desktop computer to a handheld through serial or USB cables, and it works well for all devices.

The BlackBerry 85x and 95x devices were C++ based and used an ALI file to load the required DLL files onto the handhelds. ALI files are not discussed in this hack because current model BlackBerry devices do not support them.

Create an ALX file. For current model BlackBerry devices (5000, 6000, and 7000 series) the BlackBerry Desktop Manager expects an ALX file. The purpose of the ALX file is to describe the application and list the COD files that must be loaded onto the BlackBerry device. The COD files are highly optimized binary files that contain the Java bytecode class files of your application.

The easiest way to generate an ALX file for your application is by using the JDE. On the Project menu, click Generate ALX file. This will generate a default ALX file with the values that you set up in the Project Properties dialog. You may want to customize your ALX file manually, especially if you are using an automated build process such as Ant. The ALX file is basically an XML file. Here is an example of an ALX file for a third-party BlackBerry application:

```xml
<loader version="1.0">
    <application id="IdokorroMobileAdmin">
        <name >
            Idokorro Mobile Admin
        </name>
        <description >
            Mobile Admin allows remote network administration
        </description>

        <version>3.0</version>

        <vendor >
            Idokorro Mobile Inc.
        </vendor>
        <copyright >
            Copyright (c) 2005 Idokorro Mobile Inc.
        </copyright>
        <fileset Java="1.0">
            <files >
```

```
                MobileAdminImages.cod
                MobileAdminUtil.cod
                MobileAdmin.cod
            </files>
        </fileset>
    </application>
</loader>
```

The `id` attribute of the `<application>` tag is a unique identifier used by the BlackBerry device to identify your application and to determine if the user is upgrading an existing application. On the BlackBerry handheld, users can view the ID or delete the application by selecting Options and then Applications. Pick a unique value that will not be confused with other available applications.

The `<name>` tag is the name of the application as it appears on the Home screen of a BlackBerry device.

The `<description>`, `<vendor>`, and `<copyright>` tags are used by the Black-Berry Desktop Manager to provide more information about an application.

The `<version>` tag allows you to specify the current version of your application. If you change this value after the application has been loaded onto a handheld, the Desktop Manager will recognize that an upgrade is available for the application. If you use a build script with a build system such as Ant, your script can automatically update the version number when creating new builds.

The `<fileset>` tags indicate the location of the actual COD files that the Black-Berry Desktop Manager will load. You can specify multiple COD files. It is also possible to specify attributes that will target different BlackBerry models. For more information about using these attributes, refer to the JDE documentation.

Package the application for download. After the ALX file and the COD files have been generated, the application can be packaged for download. The simplest way is to compress the required ALX and COD files in a ZIP file and provide it as a downloadable file on your web site.

Load the application using the Desktop Manager. To load a new application using the BlackBerry Desktop Manager, the user double-clicks the Application Loader icon. The Application Loader Wizard appears and displays all currently installed applications. The user clicks Add and then browses to the ALX file for your application (see Figure 9-6).

The most common problem that users experience when using the Application Loader Wizard is receiving the error message "No Additional Applications designed for your handheld were found."

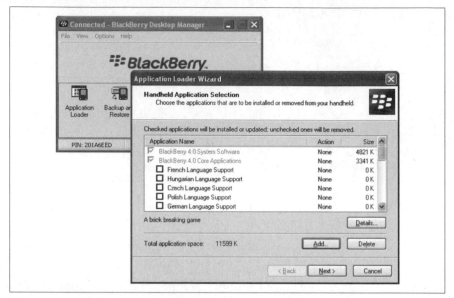

Figure 9-6. BlackBerry Application Loader

The two most common causes are:

- The Application Loader cannot find the COD file(s) because the user only copied the ALX files to the PC.
- The BlackBerry Desktop Manager does not have a local copy of the handheld operating system installed that is compatible with the device currently connected to the PC.

The latter cause is easy to diagnose by checking to see whether the user's application list is empty. If so, the user must download the BlackBerry handheld operating system for her specific device from her carrier's web site.

Deploy Applications over the Air

The best and easiest way to deploy BlackBerry applications is via over-the-air (OTA) downloads. To install an application OTA, the user navigates to the location of the application files using the BlackBerry Browser, and then installs the application directly over the wireless network. The advantage of this installation method is that a connection to a desktop PC is not required, and the download and installation can be performed anywhere.

Create a JAD file. OTA downloads require a file similar to the ALX file to identify and describe the application. A JAD file is used for this purpose. This file originates from the Java MIDP specification and is commonly used for J2ME application downloads for mobile phones. However, for Black-Berry applications, a few additional properties have been added to the JAD file format.

Here is an example of a JAD file for a BlackBerry application:

```
Manifest-Version: 1.0
MicroEdition-Configuration: CLDC-1.0
MicroEdition-Profile: MIDP-1.0
MIDlet-Version: 3.0.1
MIDlet-Name: IdokorroMobileAdmin
MIDlet-Jar-URL: IdokorroMobileAdmin.jar
MIDlet-Jar-Size: 1
MIDlet-1: Idokorro Mobile Admin,img/mobileadmin.png,admin
MIDlet-2: Idokorro Mobile Terminal,img/terminal.png,terminal
MIDlet-Vendor: Idokorro Mobile Inc.
RIM-COD-Module-Dependencies: net_rim_os,net_rim_
cldc,MobileAdminImages,MobileAdminUtil
RIM-COD-URL-1: MobileAdminImages.cod
RIM-COD-Module-Name-1: MobileAdminImages
RIM-COD-Size-1: 66004
RIM-COD-URL-2: MobileAdminUtil.cod
RIM-COD-Module-Name-2: MobileAdminUtil
RIM-COD-Size-2: 39076
RIM-COD-URL-3: MobileAdmin.cod
RIM-COD-Module-Name-3: MobileAdmin
RIM-COD-Size-3: 92896
```

The first three lines of the file indicate the minimum version of J2ME supported by the application. If your application is compatible with MIDP 1.0 and 2.0, use the lower version number.

Use the `MIDlet-Version` line to specify the version of your application. If you change this value, the user will be prompted to upgrade on subsequent downloads of the application.

Use the `MIDlet-Name` line to provide a unique name for your application. Be sure to choose a name that is unique and that will not be easily confused with other applications.

The `MIDlet-Jar-URL` and `MIDlet-Jar-Size` properties are not used by the BlackBerry device, but they must still be specified or the BlackBerry device will report the error "Invalid JAD File." In non-BlackBerry J2ME applications, OTA downloads are usually composed of a JAD file and a JAR file, which contain the Java CLASS files. These are irrelevant for BlackBerry downloads because the BlackBerry has its own special COD file format. If your application is not BlackBerry specific, you can still use a single JAR file, and the BlackBerry Enterprise Server (BES) will automatically convert the JAR file into a COD file when the user downloads the application files to a BlackBerry handheld.

The MIDlet-1 property is series of three comma-separated values (CSV). The first value is the name of the application as it appears on the Home screen of a BlackBerry device. The second value is the graphic file for the icon that will be displayed on the Home screen. The third value is for the arguments that will be passed to your public static void main(String[] args) method. Multiple application icons can be displayed simply by entering subsequent lines such as: MIDlet-2, MIDlet-3, etc.

Use the MIDlet-Vendor line to specify the company name that you want to display for your application.

Use the RIM-COD-Module-Dependencies line to specify what libraries your application depends on. For example, if it depends on the BlackBerry Browser module, the Phone module, or other third-party libraries, specify it here.

Use the RIM-COD-URL-1 line to specify the location of the COD file for your application. Use the RIM-COD-Module-Name-1 line to provide a unique name for the module. Use the RIM-COD-Size-1 line to indicate the COD file. If you have more than one COD file for the application, repeat these three lines for each COD file and increase the increment in the property name by one. For example: RIM-COD-URL-2, RIM-COD-Module-Name-2, RIM-COD-Size-2, etc.

Package the application for an OTA download. When the JAD file is complete, you can post the JAD and COD files on your web server for download. Simply upload the files to your server using FTP or whatever mechanism you usually use to post web pages to your web site.

For OTA downloads to work properly, you must set the proper MIME types for the JAD and COD files on your web server. The MIME type for a COD file is *application/vnd.rim.cod* and the MIME type for a JAD file is *text/vnd.sun.j2me.app-descriptor*. For more information about how to configure MIME types, please refer to your web server documentation.

Download an application OTA. To download your application from your web server OTA, users must simply point the BlackBerry Browser on their handheld to the URL of your JAD file. For example: *http://www.idokorro.com/TelnetSSH.jad*. Your users will then see something like the screen shown in Figure 9-7.

The user then clicks Download to transfer the application over the wireless network to the handheld. If the application has already been downloaded and a new version is available, the user will be asked if he wants to upgrade.

Figure 9-7. BlackBerry Browser OTA download

If you are going to post OTA downloads on your web server, it is highly recommended that you automatically redirect anyone using the BlackBerry Browser to your OTA download web page. To do this, configure your web server to redirect all User-Agents that contain the word "BlackBerry" to the OTA folder on your web server. This will save them the trouble of navigating through the rest of your site.

Push Applications

With the BES 4.0, it is now possible to wirelessly push applications to devices. This method is highly recommended for deploying an application to many handheld devices at once because it simplifies the installation and upgrade procedures.

To push an application to many BlackBerry devices at once, the receiving devices must be running Handheld System Software Version 4.0 or above, and have at least 16 MB of flash memory. Point the BES server to the ALX and COD files for the application, and select which users on your BES will receive the application. The BES can then install the files to the selected user devices and automatically push upgrades of the application to the handhelds whenever a new version is available. When the application is installed, users see a new icon on the Home screen of their BlackBerry handheld. For more information about this technique, see "Create a Simple Push Application" [Hack #90].

—*Paul Dumais*

HACK #98 Sign Your COD Files

To distribute your application, you first must request a certificate from RIM.

When you develop applications using the BlackBerry Simulator, everything runs smoothly. But once you deploy them to an actual handheld device, it may fail with an error message. To give RIM more central control, each application developer whose code accesses protected areas of the API must obtain security keys and sign his code before it will run on an actual Black-Berry handheld. This prevents viruses and nefarious actions on the Black-Berry platform.

The registration process is quite easy and self-explanatory. To get the exact information required to obtain a certificate, go to *http://www.blackberry.com/developers/na/java/tools/controlledAPIs.shtml*. Upon approval, you should soon receive some *.csi* executable files. Save and double-click these files on your Windows computer, and an application will appear that walks you through the process of generating a security key pair. You will be prompted to make up a new password that is used when you sign your application.

Sign Your Application

During the certificate process, you will have generated two files: *SigTool.db* and *SigTool.csk*. These files need to reside in the *bin* directory of the JDE.

Now build your application as usual. After a successful compile and build, go to the Build menu option and select Request Signatures. You will be presented with the Signature Tool, shown in Figure 9-8. The options are fairly self-explanatory, There are two you will be interested in using right away. This first is Add, which allows you to specify the COD files that need to be signed (this is where you will need to enter the password you created earlier during the certificate process). The second option, Request, sends a request to RIM to sign your COD files. The signature process occurs as the status column is updated. Upon a successful signature, the status will change to Signed. Note that all CODs that require a signature must have the status Signed.

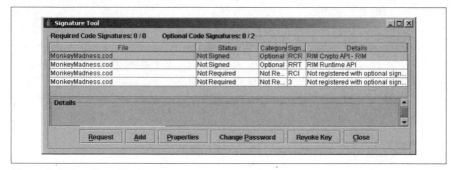

Figure 9-8. Signature Tool

Failed Signatures

There are few reasons why the signature process may fail:

- Your certificate has expired. You will need to contact RIM to extend or reapply for new certificate.

- You have exceeded the number of signatures. For security reasons, you are given a limit of how many times you can sign the COD file. But don't worry; the limit is usually some large number so it won't hinder your development. If you actually do hit the limit, you'll have to get another set of keys from RIM.

- If you reinstall or upgrade the JDE and you wipe out the SigTool files, simply drop the SigTool files back into the *bin* directory. If you lost these files you will have to contact RIM.

- The RIM Signature server may be down. Contact RIM developer support to check on this.

- Make sure you have a working Internet connection.

Once you sign your application, you can now deploy it like any other application.

Automated Build Script

Of course in the mobile world with hundreds of different devices, using a build tool such as Ant (*http://ant.apache.org/*) or Antenna (*http://antenna.sourceforge.net/*), is quite common. So the next question is whether there is a way to sign the COD files via a command-line interface (CLI). Fortunately, there is. Under the *bin* directory, you should find the executable JAR file called *SignatureTool.jar*. To use it, run java -jar SignatureTool.jar [-a] [-c] [-C] *filename*, where -a automatically request signatures, -c closes the program after a successful signing, -C forces the program to close even if the signature fails, and *filename* is of course the COD file to sign.

See Also

- *http://www.blackberry.com/developer/*
- *http://ant.apache.org/*
- *http://antenna.sourceforge.net*

—*Jason Lam*

HACK #99 Control BlackBerry Applications in the Enterprise

While the BlackBerry platform is relatively easily deployed, enterprises with more stringent security concerns or those required to meet recent regulations such as Sarbanes-Oxley must take steps to assure their BlackBerry implementation is consistent with these needs.

Research In Motion provides a comprehensive set of IT policies to control the general aspects of the device. However, the BlackBerry is increasingly being used as a platform to solve business problems. As such, companies are deploying both third-party and internally developed applications to their BlackBerry devices. This hack will discuss the application of IT policies in custom applications.

There are two primary use cases for IT policies in BlackBerry solutions:

- Provide configuration data, which can change in the future, to all devices in the enterprise.
- Hide application complexity from users.

Many wireless applications depend on server resources. These server resources may be specified by a DNS name, URL, or IP address. How does a BlackBerry administrator or BlackBerry application designer allow configuration data such as this to be easily deployed to all devices and for this data to be updated? It would be possible to include a configuration file in the deployment of an application that contains important metadata. However, what happens six months later when this data must change? IT policies provide a simple way for the BES administrator to update configuration data about a given application and have it automatically pushed to all devices. It is even possible for the BES administrator to provide certain configuration data to some user groups and different configuration data to others. This might be useful to hide complex application configuration options from users who are known to require only basic functionality.

Suppose you want to add a URL to a custom BES IT policy that will be retrieved by a custom application on devices throughout the enterprise. If the URL changes in the future, the BES admin just updates the BES setting, and all devices receive the update.

Using an IT policy to provide configuration data requires that the custom policy is properly configured on the BES and that the application properly queries the BES for the custom setting.

Configure the BES

The first step in configuring the BES is creating the *custom rule*. This is done in an entirely separate area of the BES than where policies are configured and applied. In BES 4.0, launch the BlackBerry Manager and right-click on BlackBerry Server Management. Select Properties and select the Policy Rules tab shown in Figure 9-9. To create a new rule, click the New rule button.

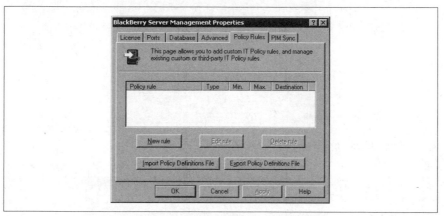

Figure 9-9. The Policy Rules tab

In this case, call the new rule CustomAppMainURL, set the type to String, and set the Destination to handheld. You must then provide a description for the IT policy. It is a common misconception that the Description field contains the value for the Rule (see Figure 9-10). The Description is just a way to document the purpose of the field; the actual default value is set later.

Rule types are not limited to String (although this is the most appropriate for this example). Other supported rule types are Boolean, integer, bitmask, and multiline string.

Now you must either edit an existing IT policy or create a new one. Create a new one as shown in Figure 9-11. Right-click on the BlackBerry server name underneath BlackBerry Server Management, select IT Policy, and click New. Create a new IT policy called ourCustomUserPolicy. Note that in addition to the standard BlackBerry policies, the new custom policy that you just created now appears.

Selecting the customAppMainURL policy immediately causes the dialog shown in Figure 9-12 to appear. Here is where you enter the actual data that you want to be sent to the custom handheld application.

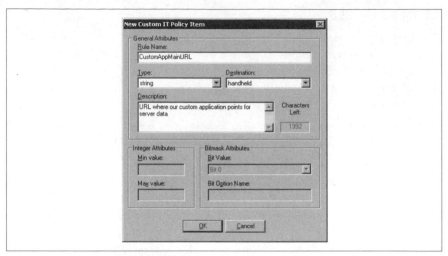

Figure 9-10. Creating the new rule

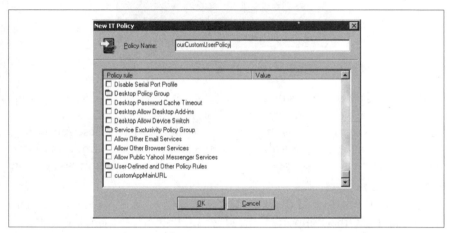

Figure 9-11. Creating the new policy

The BES server is now configured to pass down the custom data to the handheld application when asked.

 You must add the appropriate BES users to the IT policy you created or they will not receive the custom setting!

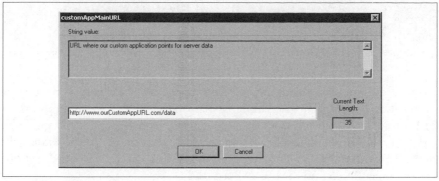

Figure 9-12. Typing the URL

Code the Custom Application

Now that the BES has been configured, all that is left is to query the BES from your application. This will not cover all aspects of building a Black-Berry application—just those points pertinent to IT policy.

In your *.java* file, include the RIM IT policy library:

```
import net.rim.device.api.itpolicy.*;
```

The code to get the IT policy is incredibly simple:

```
String itPolicyDataURL= ITPolicy.getString("customAppMainURL","http://
defaultURL");
```

Provided that you configured the IT policy correctly, the value of itPolicyDataURL will be http://www.ourCustomAppURL.com/data. In the event the device is unable to retrieve the IT policy, the http://defaultURL string that was passed as a default will be used.

There is a known bug in BlackBerry Enterprise Server 4.0 and the current 4.0 handheld software that prevents devices from properly receiving custom IT policies. The BES server will show an error on pushing an IT policy that contains any custom entries. This is clearly a serious issue that RIM is aware of. Until the bug is corrected, consider using an existing and relatively unused standard policy, and overload it to use all of your settings.

—Jeff Greenhut

HACK #100 Ensure Your App Is Placed First on the Home Screen

Make sure your program icon gets prominent placement on the Home screen after installation.

To increase the user-friendliness of your application, it's best to have your application highly visible after the installation, whether it's a desktop install or browser (OTA) installation. Ideally, you will have the application icon be the first icon on the BlackBerry Home screen (also known as the *Main Ribbon*). This commonly sought after solution and is easily overlooked in both the documentation and RIM BlackBerry JDE.

Set the Icon Position

Open your project in the BlackBerry Java Development Environment, and then right-click on it and select Properties from the context menu. Click on the Application tab and look for the label called Ribbon Position—by default, this is checked and no icon position is specified (see Figure 9-13). Uncheck the checkbox and specify the position where you want the icon. A 1 indicates position one.

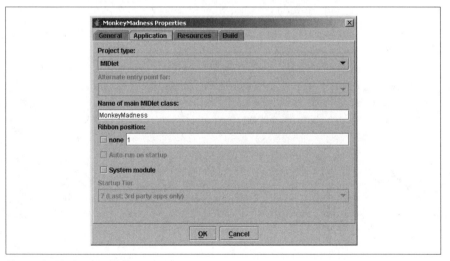

Figure 9-13. Setting icon position

That is it! You can now build and install your application, and the icon will appear in the first position. Note: you may have to clear/reset the simulator filesystem. On real devices, this is not necessary.

Position Collision

When more than one application has the icon position set to the same position, the most recently installed application will take that specified position. For example, if you install Monkey Madness first (assuming it's set to position 1), it will install, unsurprisingly, to position 1, as shown in Figure 9-14.

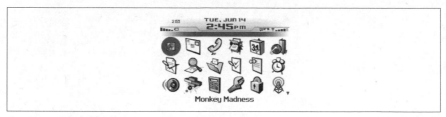

Figure 9-14. Installing first app with icon set to position 1

But if you install another app that wants position 1, such as Quotestream Wireless **[Hack #70]**, it pushes Monkey Madness over, as shown in Figure 9-15. This is only fair: every application deserves its moment in the sun!

Figure 9-15. Installing another app with icon set to position 1

—*Jason Lam*

Index

We'd like to hear your suggestions for improving our indexes. Send email to *index@oreilly.com*.

Text::CSV module, 213
themes (skins) for BlackBerry, 150–152
thermometer icons used in this
      book, xvii
thread, navigating messages in, 9
TightVNC software, 111
time, macros for, 19
T-Mobile
    BWC site for, 85
    JAD files and, 94
    service provider phone number, 23
today view
    calendar, 9
    PocketDay application, 166–169
Total Fitness for BlackBerry
      application, 197
trackwheel
    simulator mapping, 261
    using from wireless keyboard, 56
troubleshooting
    mail system not available, PIN
      messages for, 68
    messages not forwarded when
      cradled, 84
    PIN messages and, 68
typographical conventions used in this
      book, xvi

U

U shortcut, 6, 7, 8, 9
unlimited data plan, reasons to use, 2
upgrading handheld software, 52–54
uptime, determining, 10
URLs (see web pages)
USA Today web site, 138
User Admin service (see BES User
      Admin Service)
User Administration (see BlackBerry
      User Administration)
user agent, identifying browser type
      using, 241
user interface, creating in JDE, 268
user_alert.pl file, 213
users
    adding and removing, 225–229
    information about,
      exporting, 221–225
    multiple, adding to BES, 198–200
    statistics about, 225–229
    (see also owner information)
users.csv file, 222

V

V shortcut, 7, 9
VeriChat application, 122
Verizon
    BWC site for, 85
    service provider phone number, 23
View Saved Messages, shortcut for, 7
Virtual Network Computing (VNC)
      software, 111–113
Virtual Reach, Newsclip, 110
VNC client for BlackBerry, 111
VNC server, 112
VNC (Virtual Network Computing)
      software, 111–113
Vodaberry service, 36
voice emails, 173–174
voice plan, not having, 3
voicemail, showing in Messages, 8, 63

W

W shortcut, 7, 9
Wall Street Journal web site, 138
WAP Browser, launching, 7
WAP interface, accessing Gmail
      using, 79
.wav files, sending with email, 174
waypoints in map, creating, 116
weather
    searching for information about, 147
    viewing in PocketDay, 166–169
web logs
    blogging from BlackBerry, 106–109
    for BlackBerry, xv, xvi
    reading, 109–110
    spam in, avoiding, 109
web pages
    address for, 10
    bookmarking, 10, 128
    bottom of, going to, 6
    contents of, sending to
      BlackBerry, 134
    controlling access to, with
      MDS, 235–238
    CSS for layout of, 245
    delivering content based on
      browser, 240–243
    finding text in, 10
    frames in, 243
    home page, going to, 10
    images in, 244

Colophon

Our look is the result of reader comments, our own experimentation, and feedback from distribution channels. Distinctive covers complement our distinctive approach to technical topics, breathing personality and life into potentially dry subjects.

The tools on the cover of *BlackBerry Hacks* are BlackBerry mobile handhelds. Pictured on the left is the BlackBerry 7250 handheld offered by Verizon Wireless. It uses Verizon's CDMA network for voice and typically gets a data speed between 60 and 80 kbps. The phone pictured on the right is the BlackBerry 7100t offered by T-Mobile. It operates on T-Mobile's GSM network and reaches a top data speed of 40 kbps. Both handhelds offer a rich web, email, and voice experience.

Jamie Peppard was the production editor for *BlackBerry Hacks* and Linley Dolby was the copyeditor. Ann Atala proofread the book. Reba Libby and Claire Cloutier, and Darren Kelly provided quality control and Lydia Onofrei provided production assistance. Angela Howard wrote the index.

Marcia Friedman designed the cover of this book, based on a series design by Edie Freedman. The cover image is an original photograph by Marcia Friedman. Karen Montgomery produced the cover layout with Adobe InDesign CS using Adobe's Helvetica Neue and ITC Garamond fonts.

David Futato designed the interior layout. This book was converted by Keith Fahlgren to FrameMaker 5.5.6. The text font is Linotype Birka; the heading font is Adobe Helvetica Neue Condensed; and the code font is LucasFont's TheSans Mono Condensed. The illustrations that appear in the book were produced by Robert Romano, Jessamyn Read, and Lesley Borash using Macromedia FreeHand MX and Adobe Photoshop CS. This colophon was written by Jamie Peppard and Brian Jepson.

Better than e-books

Buy *BlackBerry Hacks* and access the
digital edition FREE on Safari for 45 days.

Go to www.oreilly.com/go/safarienabled
and type in coupon code JUIT-ZMRK-US32-Q9KA-QDL1

Search
thousands of
top tech books

Download
whole chapters

Cut and Paste
code examples

Find
answers fast

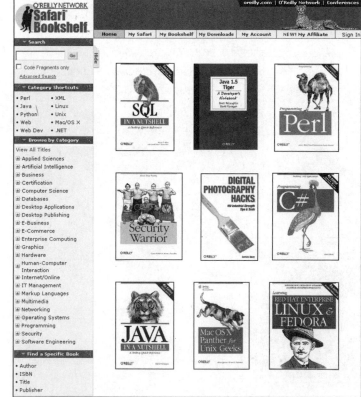

Search Safari! The premier electronic reference
library for programmers and IT professionals.

Related Titles from O'Reilly

Hardware

Blackberry Hacks

Building the Perfect PC

Car PC Hacks

Designing Embedded Hardware, *2nd Edition*

Don't Click on the Blue E!

Make: Technology on Your Time

Nokia Smartphones Hacks

Palm and Treo Hacks

PC Hardware Annoyances

PC Hardware Buyer's Guide

Smart Home Hacks

Talk Is Cheap

Treo Fan Book

Wireless Hacks

Keep in touch with O'Reilly

Download examples from our books

To find example files from a book, go to: *www.oreilly.com/catalog* select the book, and follow the "Examples" link.

Register your O'Reilly books

Register your book at *register.oreilly.com* Why register your books? Once you've registered your O'Reilly books you can:

- Win O'Reilly books, T-shirts or discount coupons in our monthly drawing.
- Get special offers available only to registered O'Reilly customers.
- Get catalogs announcing new books (US and UK only).
- Get email notification of new editions of the O'Reilly books you own.

Join our email lists

Sign up to get topic-specific email announcements of new books and conferences, special offers, and O'Reilly Network technology newsletters at:

elists.oreilly.com

It's easy to customize your free elists subscription so you'll get exactly the O'Reilly news you want.

Get the latest news, tips, and tools

www.oreilly.com

- "Top 100 Sites on the Web"—PC Magazine
- CIO Magazine's Web Business 50 Awards

Our web site contains a library of comprehensive product information (including book excerpts and tables of contents), downloadable software, background articles, interviews with technology leaders, links to relevant sites, book cover art, and more.

Work for O'Reilly

Check out our web site for current employment opportunities:

jobs.oreilly.com

Contact us

O'Reilly Media, Inc.
1005 Gravenstein Hwy North
Sebastopol, CA 95472 USA
Tel: 707-827-7000 or 800-998-9938
(6am to 5pm PST)
Fax: 707-829-0104

Contact us by email

For answers to problems regarding your order or our products:
order@oreilly.com

To request a copy of our latest catalog:
catalog@oreilly.com

For book content technical questions or corrections: **booktech@oreilly.com**

For educational, library, government, and corporate sales: **corporate@oreilly.com**

To submit new book proposals to our editors and product managers:
proposals@oreilly.com

For information about our international distributors or translation queries:
international@oreilly.com

For information about academic use of O'Reilly books:
adoption@oreilly.com
or visit:
academic.oreilly.com

For a list of our distributors outside of North America check out:
international.oreilly.com/distributors.html

Order a book online

www.oreilly.com/order_new

 **O'REILLY®**

Our books are available at most retail and online bookstores.
To order direct: 1-800-998-9938 • *order@oreilly.com* • *www.oreilly.com*
Online editions of most O'Reilly titles are available by subscription at *safari.oreilly.com*